T0271368

Tribo-Informatics

This book details how artificial intelligence and other informatic methods can be applied to the field of tribology. Using problems often found within tribological condition monitoring, behavior prediction, system optimization, and mechanism analysis, the book covers methods used such as artificial neural networks (ANN).

Including case studies throughout, the book offers an accessible introduction to tribological research, beginning with background on the theory behind tribo-informatics and updates in the latest technology. It describes how to establish a tribo-informatics database, methods through which to express tribo-systems such as ANN, support vector machines (SVM), k-nearest neighbor (KNN), and random forest (RF), and their applications. It can be used in state monitoring, behavior prediction, and system optimization. Through case studies, practical examples of how tribo-informatics can be implemented are shown throughout various industries.

This book will be of interest to students and researchers in the fields of tribology, friction, wear, and artificial intelligence.

Tribo-Informatics
The Systematic Fusion of AI and Tribology

Zhinan Zhang
Nian Yin

CRC Press
Taylor & Francis Group
Boca Raton London New York

CRC Press is an imprint of the
Taylor & Francis Group, an **informa** business

Designed cover image: Zhinan Zhang, Nian Yin, and Zhangli Hou

First edition published 2024
by CRC Press
2385 NW Executive Center Drive, Suite 320, Boca Raton FL 33431

and by CRC Press
4 Park Square, Milton Park, Abingdon, Oxon, OX14 4RN

CRC Press is an imprint of Taylor & Francis Group, LLC

© 2024 Zhinan Zhang and Nian Yin

ISBN: 978-1-032-73903-8 (hbk)
ISBN: 978-1-032-74177-2 (pbk)
ISBN: 978-1-003-46799-1 (ebk)

DOI: 10.1201/9781003467991

Typeset in Times New Roman
by MPS Limited, Dehradun

Contents

Preface

Friction refers to the resistance caused by the relative movement of objects. Humans have acknowledged and harnessed the power of friction for thousands of years. The extensive utilization of friction has greatly enhanced human productivity and quality of life. Friction mainly investigates friction, wear, and lubrication.

The tribology's multidisciplinary nature is becoming more evident along with the continuous development of relevant basic disciplines. The progress in computer science has propelled the advancement of tribology system calculations, design methods, simulation methods, and the validation of theoretical models. As information technology advances swiftly, it has given rise to effective approaches for collecting, classifying, storing, retrieving, analyzing, extracting, and disseminating research information across diverse fields and disciplines. However, due to its inherent complexity and interdisciplinary nature, the fusion of tribology and informatics presents significant challenges. Consequently, there have been constraints on the research outcomes pertaining to tribo-informatics. The architecture of tribo-informatics includes connection of the database, establishment of the database, establishment of a functional database, as well as retrieval and dissemination of information.

We wrote this book in the hope of offering some insights of tribo-informatics, covering the development of tribology and informatics, the conceptual framework and informational expression of tribo-informatics, the methods, and implementation of tribo-informatics with some real case studies.

Acknowledgments

Writing a book is a journey that is made possible by the support, encouragement, and assistance of many individuals. The authors extend heartfelt thanks to our labmates, including but not limited to Li Zhen, Wu Zishuai, He Ke, Chen Shi, Zhou Huihui, Sun Yue, Ma Yufei, Liu Songkai, Zhao Yuxin, Chen Jiawei, Lin Zemin, Hou Zhangli, Huang Kaiyi, He Zhixuan and Renaldy Dwi Nugraha. This book would not have been possible without their support.

We would like to express our sincere gratitude to the funds for their support of our research, including the National Natural Science Foundation of China (Grant Nos 12072191, 51875343, 51575340, 51205247, and U1637206) and State Key Laboratory of Mechanical System and Vibration Project (Grant Nos.MSVZD201912, MSVZD202108).

Author's biography

Zhinan Zhang

He received his Ph.D. from Shanghai Jiao Tong University. From 2011 to 2013, he worked as a postdoc at Shanghai Jiao Tong University. He is currently a professor at the School of Mechanical Engineering, Shanghai Jiao Tong University. His research interests include tribo-informatics, contact electrification, computational design, and analysis of tribo-systems. Prof. Zhang serves as the editorial board member of Friction, associate editor of ASME Journal of Computing and Information Science in Engineering, and member of international advisory board of Digital Twin.

Nian Yin

He received his B.S. degree and M.D. in mechanical engineering in 2017 and 2020, respectively, from Shanghai Jiao Tong University, Shanghai, China. He is now working as a Ph.D. student at the School of Mechanical Engineering, Shanghai Jiao Tong University. His research focuses on data-driven tribological design theories and methodologies, as well as the investigation of degradation mechanisms and protective strategies for spatial current-carrying behaviors.

1 Introduction

1.1 DEVELOPMENT OF TRIBOLOGY

Friction refers to the resistance caused by the relative movement of objects. Humans have acknowledged and harnessed the power of friction for thousands of years. During that time, two significant applications emerged: generating heat through friction and utilizing rolling friction. The extensive utilization of friction has greatly enhanced human productivity and quality of life. Tribology is defined as a field of study encompassing the theory and application of two surfaces that have relative motion and interaction [1]. Therefore, the field of friction mainly investigates friction, wear, and lubrication.

Tribology's multidisciplinary nature is becoming more evident with the continuous development of relevant basic disciplines. For instance, when integrating tribology and materials science, it is possible to utilize various materials' preparation methods [2] and performance parameters [3] to design interfaces that exhibit favorable tribological properties. The integration of physics and tribology enables the guidance of interface texture design [4,5] and the explanation of friction reduction mechanisms in 2D materials [6]. In addition, the combination of chemistry and tribology has advanced the field of tribochemistry, facilitating the study of the influence of friction processes on the chemical properties and the influence of surface chemical modification on friction properties [7,8]. The progress in computer science has propelled the advancement of tribology system calculations, design methods, and simulation methods [9,10], and the validation of theoretical models [11].

The advancement of tribology research has led to the emergence of new tribological technologies such as superlubricity and triboelectric nanogenerators. These cutting-edge technologies exhibit a multidisciplinary nature by incorporating various fields. For example, achieving super lubricity entails new materials design with knowledge of material science, surface texture optimization with knowledge of physics, and chemical surface modifications with knowledge of chemistry. Enhancing the performance of triboelectric nanogenerators involves the alterations of friction pair materials and modifications of their physical or chemical properties. The development of computer science facilitates the simulation of these technologies. However, the progress of these advancements has been hindered by ineffective tribological information exchange across various disciplines. Hence, the establishment of a tribology database becomes crucial to consolidate and integrate tribological knowledge from diverse domains, thus supporting the development of new tribology technologies more effectively.

DOI: 10.1201/9781003467991-1

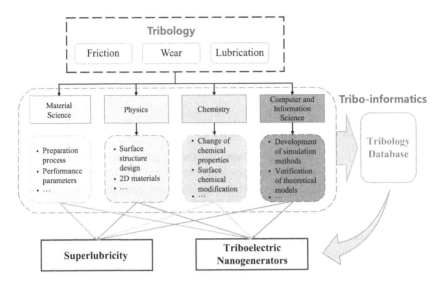

FIGURE 1.1   Illustration of current tribology research.

Figure 1.1 shows the current tribology research.

1.2   DEVELOPMENT OF INFORMATICS

With the progress of computational tools, information technology has advanced significantly and gained the capability to handle increasingly large volumes of data. Various technologies, such as artificial intelligence, machine learning, database technology, and cloud computing, are rapidly evolving to effectively manage and process information. In this era of exponentially growing data, information technology enhances the information-related abilities of human beings.

The application of information technology across various research domains has led to the establishment of diverse informatics disciplines. For example, the advancement of next-generation sequencing (NGS) technologies in bioinformatics has greatly accelerated the study of complex biological systems and serves as a technical foundation for biological research and medical applications [12]. Materials informatics, introduced in 2003, has undergone significant developments in areas such as furniture products, aerospace materials, anti-corrosion materials, and nanomaterials [13]. By effectively describing properties of materials, materials informatics provides a solid basis for the development of new materials. What is more, the rapid progress observed in chemical informatics [14], health informatics [15], music informatics [16], and safety informatics [17] exemplifies the integration of information technology with multiple fields.

As information technology advances swiftly, it has given rise to effective approaches for collecting, classifying, storing, retrieving, analyzing, extracting, and disseminating research information across diverse fields and disciplines.

However, due to its inherent complexity and interdisciplinary nature, the fusion of tribology and informatics presents significant challenges. Consequently, there have been constraints on the research outcomes pertaining to tribo-informatics.

1.3 APPLICATION OF INFORMATION TECHNOLOGY TO THE FIELD OF TRIBOLOGY

Ever since the initial acknowledgment of friction, the research of tribology has undergone four processes, as shown in Figure 1.2:

- Empirical science based on phenomena.
- Theoretical science based on simplified models.
- Computational science based on computational tools.
- Information science based on big data.

The utilization of information technology in the field of tribology encompasses three key aspects: tribology data's acquisition, analysis, and application.

Tribological information can be divided into two kinds of signals. The first kind is tribological logical signals including friction, wear, and some other tribological parameters. The second kind is derivative signals such as image, noise, vibration, temperature, and electrical signal. Figure 1.3 shows the details of two kinds of signals. These signals can also be employed to monitor operating states of a tribology system or predict tribology performances. Xue et al. [18] investigated the impact of surface structure and lubricant on friction vibration and noises in GCr15 bearing steel. Kwang-Hua [19] examined how the friction

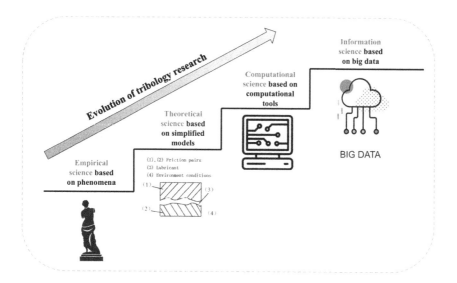

FIGURE 1.2   Four stages of tribology research.

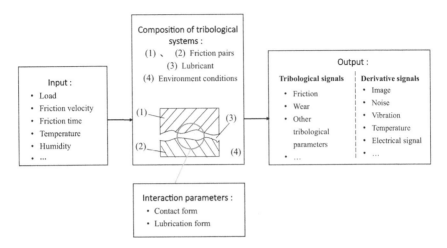

FIGURE 1.3   Input and output of tribological systems.

properties of molybdenum disulfide films with ultra-lubrication properties vary with changes in temperature.

Two kinds of signals can both be forecast by neural network methods [20–22]. The creation and implementation of a tribology database primarily involve investigating the principles of storing and retrieving various forms of tribology data. Jia et al. [23] developed a methodology for constructing a database of lubricating materials and utilized a machine-learning approach to forecast the correlation between tribology and oxidation resistance. These findings offer valuable insights for expediting the design, development, and utilization of lubricating materials. Nevertheless, due to the inherent systematic nature of tribology, the creation of a comprehensive tribology database is still pending. During the early stages of tribology research, some scholars discussed the need to establish a literature database specifically dedicated to the field of tribology [24], [25]. Nonetheless, due to a scarcity of research focusing on analyzing this particular technology, the development of such a database has not been thoroughly explored.

The ongoing integration of information technology and tribology opens up new avenues for research in the field. The establishment of a database creates favorable circumstances for the preservation and dissemination of tribology research data. The primary objective of this book is to construct a comprehensive framework for organizing and structuring tribology information in a systematic manner. Some examples will be offered to facilitate the understanding.

REFERENCES

[1] W. E. Jamison, "Introduction to tribology," *Journal of Vacuum Science and Technology*, vol. 13, no. 1, pp. 76–81, Jan. 1976, doi: 10.1116/1.568961.

[2] A. A. Anber and F. J. Kadhim, "Preparation of nanostructured SixN1−x thin films by DC reactive magnetron sputtering for tribology applications," *Silicon*, vol. 10, no. 3, pp. 821–824, May 2018, doi: 10.1007/s12633-016-9535-4.

[3] M. Milosevic, P. Valášek, and A. Ruggiero, "Tribology of natural fibers composite materials: An overview," *Lubricants*, vol. 8, no. 4, 2020, doi: 10.33 90/lubricants8040042.

[4] P. G. Grützmacher, F. J. Profito, and A. Rosenkranz, "Multi-scale surface texturing in tribology-current knowledge and future perspectives," *Lubricants*, vol. 7, no. 11, p. 95, 2019, doi: 10.3390/lubricants7110095.

[5] R. Yu and W. Chen, "Research progress and prospect of surface texturing in industrial tribology," *Jixie Gongcheng Xuebao/Journal of Mechanical Engineering*, vol. 53, no. 3, pp. 100–110, Feb. 2017, doi: 10.3901/JME.201 7.03.100.

[6] E. Cihan, S. Ipek, E. Durgun, and M. Z. Baykara, "Structural lubricity under ambient conditions," *Nature Communications*, vol. 7, Jun. 2016, doi: 10.1038/ ncomms12055.

[7] T. Demirbas and M. Z. Baykara, "Nanoscale tribology of graphene grown by chemical vapor deposition and transferred onto silicon oxide substrates," *Journal of Materials Research*, vol. 31, no. 13, pp. 1914–1923, Jul. 2016, doi: 10.1557/ jmr.2016.11.

[8] X. Liang, Z. Liu, and B. Wang, "Physic-chemical analysis for high-temperature tribology of WC-6Co against Ti–6Al–4V by pin-on-disc method," *Tribology International*, vol. 146, Jun. 2020, doi: 10.1016/j.triboint.2020.106242.

[9] D. B. Luo, V. Fridrici, and P. Kapsa, "A systematic approach for the selection of tribological coatings," *Wear*, vol. 271, no. 9–10, pp. 2132–2143, Jul. 2011, doi: 10.1016/j.wear.2010.11.049.

[10] J. K. Nørskov, T. Bligaard, J. Rossmeisl, and C. H. Christensen, "Towards the computational design of solid catalysts," *Nature Chemistry*, vol. 1, no. 1, pp. 37–46, Apr. 2009, doi: 10.1038/nchem.121.

[11] H. Hölscher, D. Ebeling, and U. D. Schwarz, "Friction at atomic-scale surface steps: Experiment and theory," *Physical Review Letters*, vol. 101, no. 24, Dec. 2008, doi: 10.1103/PhysRevLett.101.246105.

[12] B. Hwang, J. H. Lee, and D. Bang, "Single-cell RNA sequencing technologies and bioinformatics pipelines," *Experimental and Molecular Medicine*, vol. 50, no. 8, Aug. 01, 2018, doi: 10.1038/s12276-018-0071-8.

[13] S. Ramakrishna *et al.*, "Materials informatics," *Journal of Intelligent Manufacturing*, vol. 30, no. 6, pp. 2307–2326, Aug. 2019, doi: 10.1007/s1 0845-018-1392-0.

[14] R. Guha *et al.*, "The blue obelisk - Interoperability in chemical informatics," *Journal of Chemical Information and Modeling*, vol. 46, no. 3, pp. 991–998, May 2006, doi: 10.1021/ci050400b.

[15] D. Ravi *et al.*, "Deep learning for health informatics," *IEEE Journal of Biomedical and Health Informatics*, vol. 21, no. 1, pp. 4–21, Jan. 2017, doi: 10.1109/JBHI.2016.2636665.

[16] E. J. Humphrey, J. P. Bello, and Y. Lecun, "Feature learning and deep architectures: New directions for music informatics," *Journal of Intelligent Information Systems*, vol. 41, no. 3, pp. 461–481, Dec. 2013, doi: 10.1007/s1 0844-013-0248-5.

[17] P. J. H. Hu, C. Lin, and H. Chen, "User acceptance of intelligence and security informatics technology: A study of COPLINK," *Journal of the American Society*

for Information Science and Technology, vol. 56, no. 3, pp. 235–244, Feb. 2005, doi: 10.1002/asi.20124.

[18] Y. Xue, X. Shi, H. Zhou, G. Lu, and J. Zhang, "Effects of groove-textured surface combined with Sn–Ag–Cu lubricant on friction-induced vibration and noise of GCr15 bearing steel," *Tribology International*, vol. 148, Aug. 2020, doi: 10.1016/j.triboint.2020.106316.

[19] C. R. Kwang-Hua, "Temperature-dependent negative friction coefficients in superlubric molybdenum disulfide thin films," *Journal of Physics and Chemistry of Solids*, vol. 143, Aug. 2020, doi: 10.1016/j.jpcs.2020.109526.

[20] E. Šabanovič, V. Žuraulis, O. Prentkovskis, and V. Skrickij, "Identification of road-surface type using deep neural networks for friction coefficient estimation," *Sensors*, vol. 20, no. 3, pp. 1–12, Feb. 2020, doi: 10.3390/s20030612.

[21] H. Xie, Z. Wang, N. Qin, W. Du, and L. Qian, "Prediction of friction coefficients during scratch based on an integrated finite element and artificial neural network method," *Journal of Tribology*, vol. 142, no. 2, p. 021703, Feb. 2020, doi: 10.1115/1.4045013.

[22] H. Chang, P. Borghesani, and Z. Peng, "Automated assessment of gear wear mechanism and severity using mould images and convolutional neural networks," *Tribology International*, vol. 147, Jul. 2020, doi: 10.1016/j.triboint.2020.106280.

[23] D. Jia, H. Duan, S. Zhan, Y. Jin, B. Cheng, and J. Li, "Design and development of lubricating material database and research on performance prediction method of machine learning," *Scientific Reports*, vol. 9, no. 1, Dec. 2019, doi: 10.1038/s41598-019-56776-2.

[24] H. Tischer, "The BAM tribology index database: a key to the tribological literature," *Tribology International*, vol. 22, no. 2, pp. 121–124, 1989, doi: 10.1016/0301-679X(89)90172-2.

[25] J. R. Fries and F. E. Kennedy, "Bibliographic databases in tribology," *Journal of Tribology*, vol. 107, no. 3, pp. 285–294, 1985, doi: 10.1115/1.3261052.

2 Conceptual framework and informational expression of tribo-informatics

2.1 CONCEPT OF TRIBO-INFORMATICS

In order to establish a comprehensive definition of tribo-informatics, it is necessary to clarify the concepts of information entropy and tribology systems. Information entropy refers to the average amount of information remaining in a system after redundancy has been eliminated. Hence, the information entropy of a system reaches its minimum when the signal source's value is uniquely determined. The information entropy can be obtained using Equation 2.1.

$$H(U) = E\left[- \log p_i\right] = - \sum_{i=1}^{n} p_i \log p_i \qquad (2.1)$$

U is all possible results of the source value. p_i stands for the probability of the ith value in the source, and E is the expectation value of $- \log p_i$.

We take the research of super lubrication as an example in the field of tribology research. When a scholar uses "superlubricity" as the keyword to search for literature, he can find that a large number of studies have been published from 2016 to 2020. This led to an elevation in the information entropy within the field of superlubricity research, as shown in Figure 2.1. Assuming that the possibility of one particular researcher searching for a specific journal is equal to that of another researcher, the information entropy (H) is calculated as Equation 2.2 shows.

$$H = - \sum_{i=1}^{n} \frac{1}{n} \log \frac{1}{n} \qquad (2.2)$$

where n is the total number of the published papers.

Generally speaking, there are two approaches to reduce information entropy. The first one is to propose a new unified theory, which has a high demand for researchers' creativity. The second one is to systematically organize available research findings, such as the published review articles and current databases.

Figure 2.1 shows the calculation of information entropy.

DOI: 10.1201/9781003467991-2

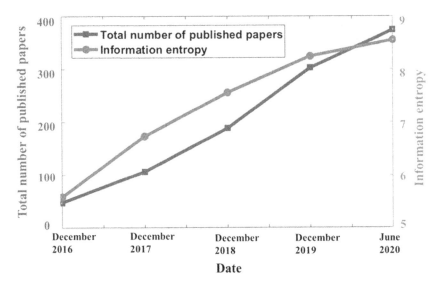

FIGURE 2.1 Calculation of information entropy.

The research on tribology is quite systematic. It mainly includes four parts: structural elements (E), element performance (P), the relationship between elements (R), and historical information (H) [1]. There are three layers of tribology systems:

- Subsystems: coatings and lubricating oils, etc.
- Current systems: friction pairs, etc.
- Super systems: bearings or other friction scenarios.

Different from bioinformatics, chemical informatics, and material informatics, tribo-informatics is mainly based on subsystem informatics. Therefore, the establishment of a tribo-informatics database is much more difficult. Figure 2.2 shows the systematic embodiment of tribology: (a) system composition and time dependence and (b) system dependence and time dependence of tribo-system.

Tribo-informatics facilitates the collection, classification, storage, retrieving, analysis, and dissemination of tribology information through establishing tribology standards, building databases, and using information technology. In this way, tribo-informatics enhances tribology research's efficiency and process.

2.2 ARCHITECTURE OF TRIBO-INFORMATICS

As shown in Figure 2.3, the architecture of tribo-informatics includes connection of the database, establishment of the database, establishment of a functional database, as well as retrieval and dissemination of information.

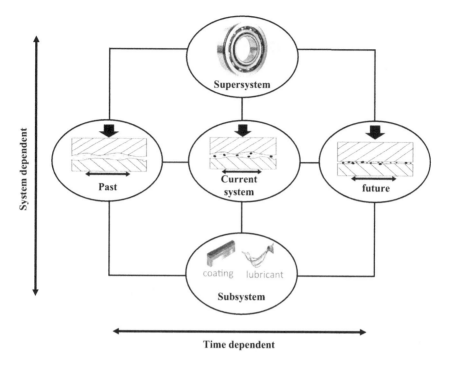

FIGURE 2.2 Systematic embodiment of tribology: (a) system composition and time dependence and (b) system dependence and time dependence of tribo-system.

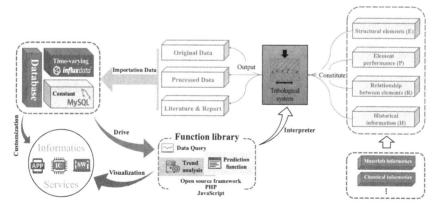

FIGURE 2.3 Architecture of tribo-informatics.

2.2.1 CONNECTION OF THE DATABASE

Tribology research is a kind of systematic science, so the input conditions of a tribology model can well be connected with databases of other disciplines, such as databases of materials, manufacturing, and chemistry.

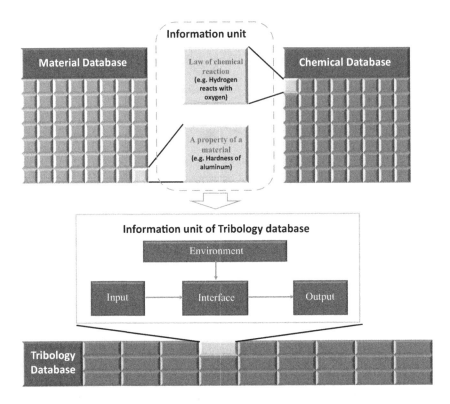

FIGURE 2.4   Information unit of tribology database.

One information unit of a tribology database is composed of four elements: input, environment, interface, and output. Figure 2.4 can serve as an example. One tribology database unit can contain a property of a certain material such as the hardness of aluminum and a law of chemical reaction such as the reaction between hydrogen and oxygen. One is from material databases, and the other is from chemical databases.

2.2.2   ESTABLISHMENT OF A DATABASE

Data can be classified into three categories according to their sources. The first category is original data, which comes from real tests following standard procedures. Therefore, this kind of data has the highest reliability. The second category is processed data. It can be obtained by theoretical prediction or simulation. The third category is literature and report data. Interestingly, literature and report data can reflect changes of hot spots in tribology fields.

Considering characteristics, the data can be treated as time-series data or relational data. Time-series data explain how the performance of a tribology system varies along with time. Relational data focuses on the relationship between output and input of a tribology system.

2.2.3 Establishment of a Functional Database

A tribology functional database can perform three functions: visualization, retrieval, and analysis, among which analysis function is the most critical one.

The analysis function is composed of a theoretical model, a simulation model, and a artificial intelligence calculation model. These models can make use of the database information to verify the system performance and to predict tribology information. Then, the prediction results can be stored as a piece of processed data. It can only be taken as a reference, since the original experience conditions cannot be achieved. Once the conditions for performing standard tests are achieved and tests are carried out, the original data should be stored.

However, the processed data will not be deleted. There are two reasons. First, the processed data are obtained as a general result based on big data, while the original data are obtained under certain conditions to rule out contingency. Second, the difference between this group of processed data and original data can be taken as reference to update the analysis functions or to infer original data based on processed data for other cases.

2.2.4 Retrieval and Dissemination of Information

Commercial applications are developed for retrieving and disseminating tribology information. The development process can refer to some mature APPs of bioinformatics and materials informatics.

People from various professions can extract different information. Scholars focusing on basic research can extract information about simulation and theoretical modes. People from industries can extract models to predict tribology systems' performance.

2.3 INFORMATIONAL EXPRESSION OF TRIBO-SYSTEM

2.3.1 Features of Tribo-System Information

Tribology behaviors are outcomes of the collaborative efforts of multiple disciplines such as mechanics, physics, chemistry, and material sciences. It clearly relies on the system and time change [2]. Therefore, it is not easy to collect and process tribo-system information considering its wide coverage and comprehensiveness characteristics.

There is a wide range of friction information sources and data structure forms, so people may find it difficult to accurately describe behaviors of a tribo-system with a single piece of physical information. Facing this challenge, it is vital to establish a systematic tribo-system information model in order to provide a tool of revealing flow laws of tribological information in various scales, levels, and among various systems. Such a model can play an important role in tribo-information collection, processing, and reuse.

2.3.1.1   Information collection

In a tribo-system, there are always two categories of performance indicators. The first type is explicit measurements, which are easy to observe, such as acoustic, electrical, vibration, and thermal measurements. The second type is implicit measurements, which are less observable, such as wear amounts, lubrication states, and surface topographies. A specific measurement can be explicit in one system, while being implicit in another system. When the critical measurements are hard to observe, status monitoring can be difficult. By taking advantage of tribo-informatics technologies, researchers are able to discover relationships between different measurements. In this way, they may infer the implicit measurements through the explicit indicators to guarantee the integrity of the tribo-system information.

2.3.1.2   Information processing

Tribological information is the result of collective effect of many disciplines, and multiple pieces of tribological information of different scales are correlated with each other. As a consequence, one single piece of tribological information contains too much information. Therefore, it is difficult to predict behaviors of tribo-systems accurately and efficiently through common physics-based analysis methods. In contrast, information technologies and methods based on machine learning and artificial intelligence are better options. They can help find the relationship among tribological information from aspects of regression, classification, clustering, and dimensionality reduction.

2.3.1.3   Information reuse

After the establishment of tribological standards and consistent data representation standards, a huge tribological information pool can be built using database technologies. The tribological database includes tribological test data, simulation data, and literature data, so that the tribological information can be reused.

2.3.2   INFORMATIONAL EXPRESSION OF TRIBO-SYSTEM

To facilitate the collection, processing, and reuse of tribological information, the tribo-system information can be divided into five categories according to the three axioms of tribology [2]: input information, system intrinsic information, output information, tribological state information, and derived state information. This can be illustrated in Figure 2.5. All information generated in working processes of any tribo-system should be included in the above five types of information.

There are four categories of tribological research, which are tribological condition monitoring, behavior prediction, system optimization, and mechanism analysis. The above five types of information are classified according to the four research categories. An essential focus of tribo-informatics research is to obtain the correlation among different pieces of information. This can be obtained

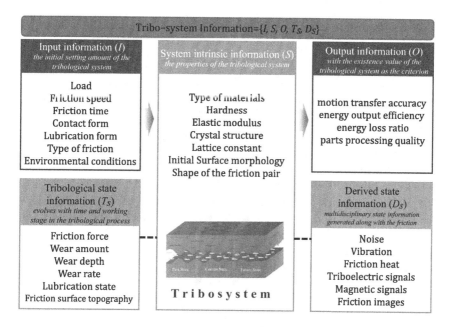

FIGURE 2.5   Information expression of tribo-system.

through informatics technologies such as machine learning and artificial intelligence.

2.3.2.1   Input information (*I*)

Input information refers to the initial settings of tribo-systems, including the load, friction speed, friction time, contact form (e.g., point-to-surface contact, surface-to-surface contact), lubrication form (e.g., dry friction, oil lubrication, grease lubrication), friction types (e.g., sliding friction, rolling friction), and environmental conditions (e.g., temperature, humidity, vacuum, radiation intensity). The input information can be determined through identification of the initial state of the work, while it is not dependent on time evolution.

Scholars can improve performances of tribo-systems by adjusting input information in research such as the study of tribo-system optimization. They can also offset negative effects of some inputs in research such as tribology under extreme operating conditions.

2.3.2.2   System intrinsic information (*S*)

System intrinsic information refers to inherent properties of the tribo system, including the surface information of the friction pair and the interface information related to the contact of the friction pair. Such information includes material types, hardness, elastic modulus, crystal structures, lattice constants, initial surface morphology, properties of the lubricant, and shapes of the friction pair.

Such information always exists, whether the system is running or not. For example, the surface information is available even when the system is not working, so it can be recognized as system intrinsic information. However, the friction force does not exist without the relative motions, so information describing the friction force is not system intrinsic information.

The information might change as the system operates, such as surface topography. However, the eigenvalues usually refer to initial values, so the intrinsic parameters of the tribo-system should be confirmed at the beginning of the work.

2.3.2.3   Output information (O)

The output information depends on the value of the tribo-system and reflects how the system achieves designed functions. Key functions of the tribo-system include motion transfer, energy transfer, information transfer, and material processing. Considering this, the output information of a tribo-system includes the accuracy degree of motion transfer, the efficiency of energy output, the ratio of energy loss, and the quality of processed materials. Output information serves as an important indicator for evaluating the working performance and predicting remaining life of the tribo-system. For example, researchers sometimes take the wear amount as the evaluation reference to predict remaining life, because the wear reduces accuracy of motion transmission and improves the proportion of energy loss.

Therefore, to accurately identify the output information, it is important to analyze the value of the tribo-system first.

2.3.2.4   Tribological state information (T_s)

Tribological state information refers to the values that evolve along with time and different working stages throughout the whole tribological process. Examples include the friction force, wear amount, wear depth, wear rate, friction surface topography, and lubrication state. Tribological state information has clear characteristics of time series, and it is closely related to the input information and system intrinsic information. Meanwhile, it affects the output performance of the system, and how it achieves target functions. In traditional tribological research, tribological state information receives the most attention, and it serves as the most critical source when analyzing tribo-system behaviors.

2.3.2.5   Derived state information (D_s)

Since the tribo-system is operating under the collaborative effects of multiple disciplines, it will generate multidisciplinary state information with tribological behaviors during working processes. Such information is called derived state information. It is distributed in a wide variety of friction-derived phenomena, and the variety increases along with the further deepening of tribological research. Derived state information includes friction images, noises, vibration, friction heat, triboelectric signals, and magnetic signals.

Such information is closely related to tribological state information, which can be explicit or implicit. For example, image information can reflect friction surface morphology. Vibration information is related to friction force information. Therefore, the establishment of the relationship between two types of state information can enhance researchers' understanding of the working state of a tribo-system.

According to the above classification, we can obtain a general information expression equation for any tribology research object:

$$\text{Tribo} - \text{system Information} = \{I, \ S, \ O, \ T_s, \ D_s\}$$

The information can be stored accordingly. Then, researchers can associate data of different categories with informatics methods. In other words, one of the most important focuses of "tribo-informatics" research is to establish the correlation of multiple pieces of tribological information.

2.3.3   A CASE STUDY OF TRIBO-SYSTEM INFORMATION EXPRESSION

We would like to offer an example showing how to apply information expression of tribo-systems using a triboelectric nanogenerator (TENG).

TENG is an energy harvesting and output device based on friction. It is now widely applied in various fields such as sensing and energy supply. Tribo-system information expression can offer a more complete understanding of TENG by analyzing its characteristics and research purposes.

To identify the output information of a tribo-system, it is necessary to analyze the major design purpose of the system. Therefore, we probe into the purpose of TENG first. The key functions of TENG are sensing and energy supply, so the major purpose of TENG is to transfer information and energy.

Then, we discuss TENG based on purposes of typical tribological research. As mentioned above, the purposes of tribological research include state monitoring, behavior prediction, system optimization, and mechanism analysis. How are these concepts applied in the research of TENG?

- **State monitoring:** In research related to TENG, state monitoring covers tribological and derived state information such as output current, voltage, friction force, and wear amount.
- **Behavior prediction:** This concept focuses on the evolution of the same group of values along with time.
- **System optimization:** Since the main purpose of TENG is to transfer information and energy, the key measures of TENG performance should be indicators such as information transfer efficiency, information carrying capacity, output functions, and power conversion efficiency. By adjusting the input information (e.g., relative motion speed and vibration frequency) and the system intrinsic information (e.g., surface contact materials), researchers can optimize the system performance.

The utilization of the informational expression model of the tribo-system can offer a more comprehensive understanding to researchers and help improve research efficiency.

REFERENCES

[1] H. Czichos, "A systems analysis data sheet for friction and wear tests and an outline for simulative testing," *Wear*, vol. 41, no. 1, pp. 45–55, 1977, doi: 10.1016/0043-1648(77)90190-9.
[2] Y. B. Xie, "On the tribology design," *Tribology International*, vol. 32, no. 7, pp. 351–358, Jul. 1999, doi: 10.1016/S0301-679X(99)00059-6.

3 Methods and implementation of tribo-informatics

3.1 CONNOTATION OF TRIBO-INFORMATICS METHODS

Tribo-informatics methods refer to all methods to process tribology information. They mainly include traditional information processing methods and advanced machine learning methods. Traditional information processing methods include linear regression models, least square methods, and Gaussian regression models. Machine learning methods include artificial intelligence (AI) methods. The major purpose of applying tribo-informatics methods is to analyze the relationship among various values of the tribo-system:

- **State monitoring:** To monitor the status of the tribo-system, researchers need to obtain the relationship between observable state values and unobservable state values.
- **Behavior prediction:** To predict system behaviors, researches need to obtain the relationship between the future state values and the input information or the current state values.
- **System optimization:** To optimize the system performance and get a better output, researchers need to figure out the relationship between the output and the input.

As shown in Figure 3.1, the application of tribo-informatics can help in regression, classification, clustering, and dimensionality reduction. Regression models can determine the quantitative tribological relationship. Classification methods determine characteristics of tribological behaviors. Clustering methods discover the new laws of tribology. Dimensionality reduction methods can improve research efficiency of tribology.

The most commonly used methods among these are artificial neural networks (ANN), support vector machines (SVM), k-nearest neighbor (KNN), and random forest (RF).

3.1.1 ANN IN TRIBOLOGY RESEARCH

ANN consists of a large number of node connections. Every node represents an incentive function, and the connection between two nodes represents a weight [1].

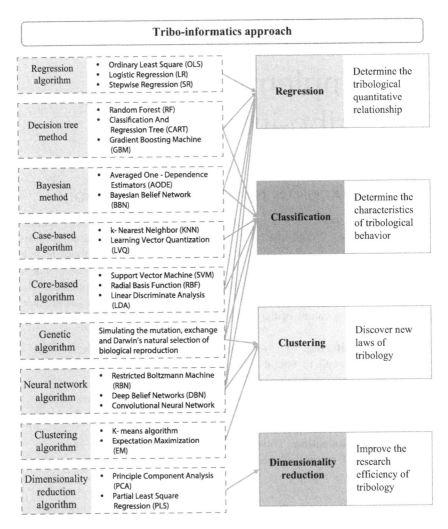

FIGURE 3.1 Classification and function of tribo-informatics methods.

It has non-linear characteristics, and it plays an important role in regression, classification, and clustering. ANN has three main types of units: input unit, hidden unit, and output unit.

- **Input unit:** It accepts signals and data from the outside world. In tribological research, the system input information such as friction speed and load can be seen as input units.
- **Output unit:** It realizes the output of the processing result of the system. The system output information such as friction, wear, and lubrication can be seen as output units.

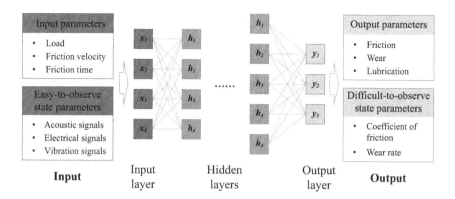

FIGURE 3.2   Application of ANN methods in tribology research.

- **Hidden unit:** It is located between the input and output units, and it cannot be observed.

Interestingly, some observable values, such as acoustic signals and electrical signals, can be used as input units to infer the unobservable values, such as friction coefficients and wear rates. Figure 3.2 shows the application of ANN methods in tribology research.

3.1.2   SVM in Tribology Research

SVM is a generalized linear classifier that classifies binary data according to supervised learning [2]. It aims to find the hyperplane that is the farthest away from various sample points (i.e., the hyperplane with the most significant separation). When the hyperplane and sample points cannot be entirely linearly separated, slack variables are introduced. When they are not linearly separable, the sample points are mapped to a high-dimensional space to be linearly separable. SVM is supported strictly in mathematics and has strong interpretability. Therefore, it is often applied to behavior classification in tribological research. Figure 3.3 shows the application of SVM methods in tribology research.

3.1.3   KNN in Tribology Research

KNN is one of the simplest classification algorithms, and it is the most commonly used [3]. It uses K nearest categories to determine which category it belongs to. The determination of K is important for classification. The error rate decreases first and then increases along with the increase of K value. Therefore, Figure 3.4 can serve as an example that there is a critical optimal value of K when determining a batch of samples.

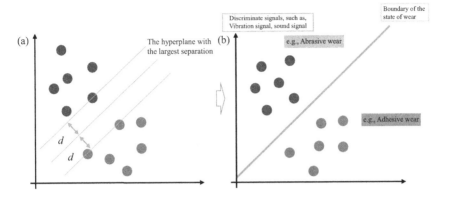

FIGURE 3.3 Application of SVM methods in tribology research: (a) Principle of SVM; (b) Application of SVM in the classification of tribological behavior characteristics.

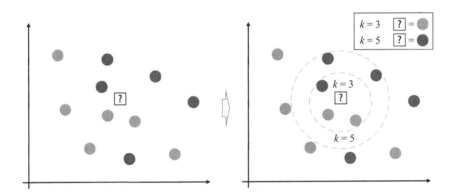

FIGURE 3.4 Simplified schematic of KNN.

In tribological research, KNN is usually used to predict the coefficient of friction (COF) and wear rates [4].

3.1.4 RF in Tribology Research

RF is a classifier containing multiple decision makers, and this method can be used for classification and regression, as shown in Figure 3.5. It can achieve a high prediction accuracy using a small amount of calculation. Meanwhile, it is not sensitive to the lack of some data parts.

In the RF method, researchers perform random sampling with replacement based on the original training set to obtain K training sets. After that, researchers select m features for each training set randomly to form K classification models. Finally, they vote for the best classification. In this way, the number of decision trees is increased so that the shortcomings of a single decision tree are overcome.

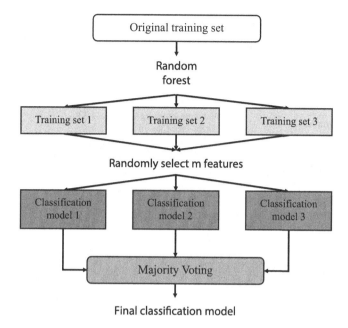

FIGURE 3.5 Simplified schematic of RF.

The application of the RF method in tribological research can facilitate the prediction of tool wear [5,6], COF, and wear rate [4,7].

3.2 APPLICATION OF TRIBO-INFORMATICS METHODS IN TRIBOLOGY

From the perspective of tribological research's purpose, the application of tribo-informatics methods can be divided into three types: status monitoring, behavior prediction, and system optimization. This section will discuss the operation procedures targeting each application purpose. Meanwhile, a framework will be offered explaining the application of tribo-informatics methods in tribology research.

3.2.1 STATE MONITORING OF TRIBOLOGICAL SYSTEM

The state monitoring of a tribo-system plays an important role in the real-time diagnosis of faults and maintenance of the system, as shown in Figure 3.6. It can be achieved by using the by-product information to monitor the system state. For example, the working state of the bearing can be determined by the state of the friction force and the friction torque of the bearing [8,9] using methods such as RF, gradient boosting classifier (GBC), and extra tree classifier (ETC). In this way, the state of unobservable values such as friction and lubrication can be

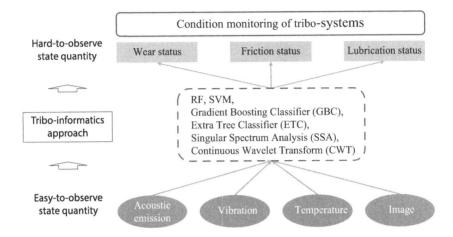

FIGURE 3.6 Application of tribo-informatics approach in tribological condition monitoring.

monitored through the track of observable state values such as image, temperature, sound pressure, lubricating oil quality, and vibration.

3.2.1.1 Wear status monitoring

Wear is a factor affecting the performance of a mechanical system the most directly. The change of wear state can lead to abnormalities in sound signals [10], image signals, and vibration signals [11]. The state of wear behavior can also be obtained by analyzing the abrasive particles. Wear state monitoring has two main categories: machined parts wear monitoring and functional parts wear monitoring.

3.2.1.1.1 Machined parts wear monitoring

A typical application of wear state monitoring is tool wear monitoring, which greatly affects the processing quality and efficiency. Engineers can identify the correct time for repairing or changing tools if the tools' states are monitored in real time.

There are different types of information related to the tool's wear, including sound pressure, images, and vibration signals. Multiple technologies and algorithms can be adopted to facilitate the tool's wear state monitoring. The SVM learning model can find the relationship between the phenomenon of tool wear and sound domain signals [12]. Expanding regularized particle filtering technique is called the semi-physical model, which can reduce errors due to pure data processing [13]. Machine learning methods integrating multiple features and multiple models can monitor tool wear with a dynamic smoothing scheme [14]. Pseudo-local singular spectrum analysis (SSA) can also monitor tool wear states by processing vibration signals [15].

Currently, most research related to tool wear state monitoring is limited in a qualitative level. The implementation of decision-making algorithms can further enhance the efficiency.

3.2.1.1.2   Functional parts wear monitoring

Wear behaviors can be observed in movable connection parts in mechanical systems, so wear state monitoring can offer a reference for the mechanical system's operating quality and remaining life. Chang et al. proposed a method of evaluating the degree of wear referring to image data sets [22]. Acoustic emission is also a critical technology for wear state monitoring. It helps to distinguish wear states into running-in, inadequate lubrication, and particle-contaminated oil [16,17]. Using continuous wavelet transform (CWT) and SVM as classifiers is an effective method to monitor wear status [18]. What is more, shape characteristics of wear particles can also reflect wear states. The radial concave deviation (RCD) method can analyze the relationship between wear states and the structure of wear particles [19,20].

3.2.1.1.3   Friction status monitoring

In the available tribology research, most friction state monitoring is qualitative, and it is used to define the deterioration of friction [18,21] mostly. The research on friction state monitoring is not sufficient.

However, the state monitoring of frictional forces and frictional torques is important to mechanical systems that require high precision and stability, such as aero-engine shaft-bush adjustment system and satellite attitude control system (e.g., momentum wheel). Friction state monitoring is playing a more and more significant role as industrial products have a higher demand for precision and reliability.

3.2.1.1.4   Lubrication status monitoring

Lubrication is a key factor affecting wear and friction. Considering this, most of the current lubrication state monitoring is carried out from two perspectives, which are friction status [22] and oil status [23]. ANN and LDA can help classify the lubricants in various states.

3.2.2   TRIBOLOGICAL SYSTEM BEHAVIOR PREDICTION

Behavior prediction of a tribological system can be greatly achieved by integrating information technologies and tribology. It is significant for predicting failure and remaining service life of mechanical systems.

Behavior prediction of tribological systems is usually applied in the following scenarios: cutting processing [24], friction stir welding [25], geological tribology [26–28], and basic tribological research. As shown in Figure 3.7, this section will discuss from three perspectives: wear prediction, friction prediction, and lubrication prediction.

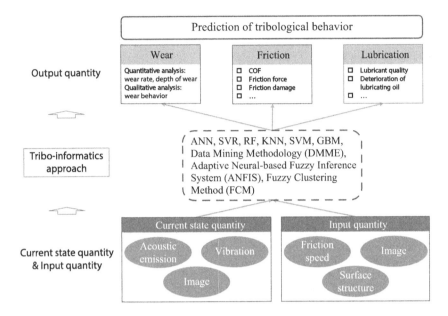

FIGURE 3.7   Application of tribo-informatics approach in the prediction of tribological behavior.

3.2.2.1   Wear prediction

In the study related to tribo-systems, damage on mechanical parts caused by wear draws much attention from researchers. Wear prediction includes two major categories: quantitative analysis of wear and classification of wear behaviors. Quantitative analysis can be used to predict wear rates and wear amounts. Behavior prediction is to predict the behavior characteristics of wear.

3.2.2.1.1   Quantitative analysis of wear

The quantitative prediction of tool wear usually follows these steps. The first step is to collect a large amount of data. Then, the dependence of wear amount on time should be identified [29–31]. During this process, various information technologies can be used, such as ANNs [32–35], support vector regression (SVR) [36,37], adapted data mining methodology (DMME) [38,39], and RF methods [5,6]. AI and machine learning can contribute to predicting wear. Generally, researchers can learn from data using neural network methods, find expression equation of wear amounts using genetic programming, and impose rules with physical meaning using fuzzy inference systems [40].

A great portion of research on wear is performed based on the wear of tools [14,32,41]. In such research, machine learning methods are usually adopted to process test data to predict wear rates [42,43]. Part of the quantitative wear analysis studies impacts of various materials on wear. For example, the machine learning method can help study the influence of composite materials [44,45], biological friction materials [46], and materials with different surface textures on the amount of wear. Specifically, Hasan et al. compared five different machine learning algorithms [4,7], covering KNN, SVM, ANN, RF, and GBM. Their research can be taken as the reference when studying the implementation of machine learning methods in tribological problems.

3.2.2.1.2 Classification of wear behaviors

Wear behaviors can be classified according to their influencing factors, such as material processing parameters, working conditions, lubrication conditions [47,48], and material composition [49]. Many different methods can be used to predict wear behaviors, such as clustering, decision-making methods, fuzzy inference, ANN approaches, adaptive neural-based fuzzy inference system (ANFIS) technique, and fuzzy clustering method (FCM) [50].

The determination of wear states should use physical rules when determining wear states. Therefore, rule models such as ANN models, rule-based belief inference (BRB) models, and evidential reasoning (ER) can facilitate identifying wear problems [51].

3.2.2.2 Friction prediction

In cutting work, information technologies can forecast cutting forces, which are related to cutting parameters, tool geometries, and tool wear states. Therefore, the prediction of friction is closely related to state monitoring. In the state monitoring system, parameters are input into the prediction model. Then, the output can be calculated through the DNN method [24] and the SVR method [52].

In addition to friction prediction of tools, friction prediction also includes friction damage [53] and friction prediction during machining processes [54,55].

What is more, the prediction of coefficients of friction (COF) is of great importance to tribo-system operation. Some typical examples of COF are COF of coatings [21] [38,56], COF of automobiles [20], and COF of pipes [57–59]. COF is one of the most important indicators of tribology in measuring tribological components. The accurate COF prediction can help in predicting the future functional state of a tribological system and guide the selection of pre-preparation processes. Specifically, the DNN method plays an essential role in predicting COF [4,7,42].

3.2.2.3 Lubrication prediction

Prediction of lubrication performance is vital to the optimization of tribological systems. Prediction of lubrication performance analyzes the relationship between surface lubrication performance and surface characteristics, such as roughness and surface texture [60,61]. It also analyzes evolution trends of lubricant performance [62] and predicts remaining life [63].

3.2.3 TRIBOLOGICAL SYSTEM OPTIMIZATION

Optimization of a tribological system is a method of improving operational performance of this tribo-system based on prediction of its tribological

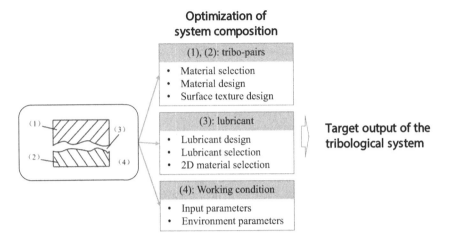

FIGURE 3.8 Optimization of the tribological system.

behaviors, which can be illustrated in Figure 3.8. It begins with the analysis of tribo-system's composition. Optimization of a tribo-system can be achieved through three ways: optimization of tribo-pairs materials, optimization of lubricants, and optimization of working conditions.

In the optimization process, AI and machine learning methods contribute much to improving design efficiency and reducing design costs.

3.2.3.1   The optimization of tribo-pairs

The optimization of tribo-pairs can be obtained in three ways, which are surface texture design, material design, and material selection.

3.2.3.1.1   Surface texture optimization

Under different surface texture conditions, AI methods can evaluate the impacts of manufacturing processes on friction by measuring friction results. In this way, researchers can control friction phenomena by designing surface textures with various parameters [64–66]. When the manufacturing process or surface texture design is different from the expectation, the tribo-system can be optimized [4], [67].

3.2.3.1.2   Surface material optimization

The design of materials can be achieved in two ways. The first is to change composition of materials, and the second is to change material structures by modifying process parameters [68]. The design of composite materials [69–71] and parameters of carburizing and nitriding process [72] can both be optimized through ANN methods.

3.2.3.1.3   Surface material selection

Compared to designing new materials of tribo-pairs, it is more common to select suitable materials for existing materials. Engineers can rely on their expertise to screen out appropriate materials or get help from information technologies. Luo et al. established a database of tribological coatings, evaluated their performances by simple tests, and matched them with various engineering needs [9]. The decision tree model can facilitate the high-throughput screening of materials [73,74].

3.2.3.2 The optimization of the lubricant

Lubricants reduce friction and wear, so designing or selecting appropriate lubricants can help optimize performance of tribological systems as well. The optimization of lubricants can be completed by synthesis of new lubricants [23,75] and allocation of lubricant ratios [76,77]. With the development of 2D materials, it is also effective to screen suitable two-dimensional materials to optimize lubrication materials.

3.2.3.3 The design of working conditions

The outputs of tribo-system are affected by input parameters (e.g., sliding speeds and loads) [78,79] and environmental conditions (e.g., temperature and humidity) [80,81]. It is effective to design suitable working conditions to optimize the performance of tribo-systems.

3.3 FUTURE TRENDS OF TRIBO-INFORMATICS

Tribo-informatics emerged in response to the rapid advancement of informatics technology. The progression of machine learning, AI, and related technologies has significantly enhanced the effectiveness of collecting, processing, analyzing, and reutilizing data. Guided by the principles of tribo-system informatization and the ideology of tribo-informatics research, it promises to infuse new momentum into tribological exploration. The future trend of tribo-informatics revolves around enhancing the acquisition and categorization of tribo-system data while expanding its conceptual essence, application contexts, and technological implementation avenues. To provide a more detailed insight into the pivotal role of tribo-informatics within tribology, forthcoming research directions will be delineated across six distinct domains, each rooted in the diverse realms of tribological investigation.

3.3.1 Basic Theories of Tribology

Foundational theoretical research exhibits a propensity for ingenuity, whereas AI technology fundamentally revolves around uncovering data linkages among physical parameters, thus precluding the direct generation of theoretical knowledge. Nonetheless, harnessing AI technology can furnish a more intuitive data groundwork for dissecting mechanisms. For instance, investigators can initially establish associations between the dynamic contact resistance of current-carrying friction pairs and signals like acoustics, vibrations, and acoustic emissions, engaging in correlation examinations. Subsequently, they can pinpoint the predominant factors influencing the performance of current-carrying friction and unveil underlying mechanisms. Conversely, the primary objective in most tribological inquiries is achieving super-slip or ultra-low wear, a pursuit of paramount importance in curtailing energy consumption through reduced friction and extending the lifespan of mechanical systems. By employing prediction methods rooted in tribo-informatics, it becomes feasible to establish the interplay between diverse inputs such as structures, compositions, environments, or inherent system characteristics and friction/wear outcomes. Consequently, a comprehensive understanding of the multitude of elements influencing super-slip/ultra-low wear can be attained, enabling the creation of a high-throughput screening framework for input/system parameters and facilitating the design of super-slip/ultra-low wear tribo-systems.

3.3.2 INTELLIGENT TRIBOLOGY

Intelligent tribology epitomizes a quintessential research avenue characterized by the principles of "tribo-informatics," and it stands as one of the most extensively explored domains. This area not only facilitates the continuous monitoring and diagnosis of tribological systems but also empowers the efficient and smart formulation of frictional and lubrication materials. One particular developmental concept warrants attention—an agile approach to controlling interfacial friction and lubrication behavior, which has witnessed a rapid evolution in recent years. This primarily encompasses real-time monitoring of coating conditions, self-repair mechanisms for damage mitigation, and proactive lubrication regulation. Within this spectrum, real-time monitoring technology serves to vigilantly assess coating conditions by establishing connections between structural variables of the coating (e.g., morphology, fissures, and wear) and readily observable derived status indicators. Damage self-repair entails a high-throughput screening of numerous lubricating materials, while active lubrication regulation necessitates the establishment of links between lubricant quantities and tribological performance. These intelligent functions encompassing detection, regulation, and restoration can establish data correlations through the tools of tribo-informatics, laying the foundation for informed decision-making processes.

3.3.3 COMPONENT TRIBOLOGY AND SURFACE TECHNOLOGY

The core elements of tribology represent crucial constituents within machinery systems, capable of independent existence and sale. The integration of AI technology in this domain should primarily emphasize the refinement of product designs and the anticipation of service performance outcomes. As an illustration, employing AI techniques to enhance the pore architecture of bearing cages can enhance the passive oil replenishment capabilities of bearings. Concurrently, forging connections between varying oil replenishment levels and fluctuations in friction torque can provide guidance on the appropriate lubrication oil volume to be applied. Enhancing the efficiency of product design and forecasting patterns of performance degradation hold paramount significance in the application of AI technology within this sphere.

3.3.4 EXTREME TRIBOLOGY

Challenging tribological issues in extreme operational settings frequently manifest in remote space, deep underground, deep-sea, polar terrains, and analogous environments. These tribo-systems necessitate exceptional precision, intricate configurations, and exceedingly elevated expenses, rendering performance and longevity predictions through iterative trials a formidable endeavor. Tribo-informatics methodologies offer a viable approach by initially amalgamating available data for sample augmentation, followed by data correlation via condition monitoring and longevity forecasting techniques. This strategy effectively mitigates the risk of imprecise predictions arising from the formidable task of directly measuring certain state parameters in the midst of extreme working conditions.

3.3.5 BIO-TRIBOLOGY

Bio-tribology primarily delves into the examination of tribological characteristics concerning joints, implants, and wearable gadgets within living organisms. Notably, the investigation of tribological performance exhibits minimal disparities between natural organism joints and internal implants when juxtaposed with tribological exploration within mechanical systems. Concerning wearable devices, especially in light of the burgeoning "Metaverse" concept, greater emphasis should be directed toward the development of tribological devices capable of environmental awareness and human–computer information interaction. This encompasses innovations like sensory artificial limbs capable of detecting movement upon contact and tactile skin rooted in tribology principles, among others. Within this domain, the pivotal application area for AI technology lies in establishing data correlations among environmental data, contact point tribological data, and cerebral signals.

3.3.6 GREEN TRIBOLOGY

Green tribology primarily investigates a range of facets, encompassing the regulation of frictional emissions, dampening friction-induced noise, the advancement of eco-friendly lubricants, and the creation of extended-life frictional interfaces. The central pursuit in green tribology centers on governing the derivatives arising from the frictional processes. AI technology plays a pivotal role in advancing green tribology by establishing data linkages between input parameters and the derivatives within tribological systems, encompassing factors such as noise levels, particle debris, and lubricant consumption.

REFERENCES

[1] O. I. Abiodun, A. Jantan, A. E. Omolara, K. V. Dada, N. A. E. Mohamed, and H. Arshad, "State-of-the-art in artificial neural network applications: A survey," *Heliyon*, vol. 4, no. 11, p. e00938, Nov. 2018, doi: 10.1016/J.HELIYON.2018. E00938.

[2] B. Schölkopf, "SVMs - A practical consequence of learning theory," *IEEE Intelligent Systems and Their Applications*, vol. 13, no. 4, pp. 18–21, Jul. 1998, doi: 10.1109/5254.708428.

[3] Z. Zhang, "Introduction to machine learning: k-nearest neighbors," *Annals of Translational Medicine*, vol. 4, no. 11, Jun. 2016, doi: 10.21037/ATM.2016. 03.37.

[4] M. S. Hasan, A. Kordijazi, P. K. Rohatgi, and M. Nosonovsky, "Triboinformatic modeling of dry friction and wear of aluminum base alloys using machine learning algorithms," *Tribology International*, vol. 161, p. 107065, Sep. 2021, doi: 10.1016/J.TRIBOINT.2021.107065.

[5] D. Wu, C. Jennings, J. Terpenny, R. X. Gao, and S. Kumara, "A comparative study on machine learning algorithms for smart manufacturing: Tool wear prediction using random forests," *Journal of Manufacturing Science and Engineering*, vol. 139, no. 7, p. 071018, Jul. 2017, doi: 10.1115/1.4036350.

[6] D. Wu, C. Jennings, J. Terpenny, S. Kumara, and R. X. Gao, "Cloud-based parallel machine learning for tool wear prediction," *Journal of Manufacturing Science and Engineering*, vol. 140, no. 4, p. 041005, Apr. 2018, doi: 10.1115/1. 4038002.

[7] M. S. Hasan, A. Kordijazi, P. K. Rohatgi, and M. Nosonovsky, "Triboinformatics approach for friction and wear prediction of Al-graphite composites using machine learning methods," *Journal of Tribology*, vol. 144, no. 1, p. 011701, Jan. 2022, doi: 10.1115/1.4050525.

[8] S. Patil and V. Phalle, "Fault detection of anti-friction bearing using ensemble machine learning methods," *International Journal of Engineering*, vol. 31, no. 11, pp. 1972–1981, Nov. 2018.

[9] V. N. Balavignesh, B. Gundepudi, G. R. Sabareesh, and I. Vamsi, "Comparison of conventional method of fault determination with data-driven approach for ball bearings in a wind turbine gearbox," *International Journal of Performability Engineering*, vol. 14, no. 3, pp. 397–412, Mar. 2018, doi: 10.23940/IJPE.18.03. P1.397412.

[10] V. Pandiyan, J. Prost, G. Vorlaufer, M. Varga, and K. Wasmer, "Identification of abnormal tribological regimes using a microphone and semi-supervised machine-learning algorithm," *Friction*, pp. 1–14, Jun. 2021, doi: 10.1007/S4 0544-021-0518-0.

[11] M. A. L. Cabral *et al.*, "Automated classification of tribological faults of alternative systems with the use of unsupervised artificial neural networks," *Journal of Computational and Theoretical Nanoscience*, vol. 16, no. 7, pp. 2644–2659, 2019, doi: 10.1166/JCTN.2019.8152.

[12] A. Kothuru, S. P. Nooka, and R. Liu, "Application of audible sound signals for tool wear monitoring using machine learning techniques in end milling," *The International Journal of Advanced Manufacturing Technology*, vol. 95, no. 9, pp. 3797–3808, Dec. 2017, doi: 10.1007/S00170-017-1460-1.

[13] H. Hanachi, W. Yu, I. Y. Kim, J. Liu, and C. K. Mechefske, "Hybrid data-driven physics-based model fusion framework for tool wear prediction," *The*

International Journal of Advanced Manufacturing Technology, vol. 101, no. 9, pp. 2861–2872, Dec. 2018, doi: 10.1007/S00170-018-3157-5.

[14] Y. Shen *et al.*, "Predicting tool wear size across multi-cutting conditions using advanced machine learning techniques," *Journal of Intelligent Manufacturing*, vol. 32, no. 6, pp. 1753–1766, Jul. 2020, doi: 10.1007/S10845-020-01625-7.

[15] B. Kilundu, P. Dehombreux, and X. Chiementin, "Tool wear monitoring by machine learning techniques and singular spectrum analysis," *Mechanical Systems and Signal Processing*, vol. 25, no. 1, pp. 400–415, Jan. 2011, doi: 10.1016/J.YMSSP.2010.07.014.

[16] F. König, J. Marheineke, G. Jacobs, C. Sous, M. J. Zuo, and Z. Tian, "Data-driven wear monitoring for sliding bearings using acoustic emission signals and long short-term memory neural networks," *Wear*, vol. 476, p. 203616, Jul. 2021, doi: 10.1016/J.WEAR.2021.203616.

[17] F. König, C. Sous, A. Ouald Chaib, and G. Jacobs, "Machine learning based anomaly detection and classification of acoustic emission events for wear monitoring in sliding bearing systems," *Tribology International*, vol. 155, p. 106811, Mar. 2021, doi: 10.1016/J.TRIBOINT.2020.106811.

[18] N. Mokhtari, J. G. Pelham, S. Nowoisky, J.-L. Bote-Garcia, and C. Gühmann, "Friction and wear monitoring methods for journal bearings of geared turbofans based on acoustic emission signals and machine learning," *Lubricants*, vol. 8, no. 3, p. 29, Mar. 2020, doi: 10.3390/LUBRICANTS8030029.

[19] W. Yuan, K. S. Chin, M. Hua, G. Dong, and C. Wang, "Shape classification of wear particles by image boundary analysis using machine learning algorithms," *Mechanical Systems and Signal Processing*, vol. 72–73, pp. 346–358, May 2016, doi: 10.1016/J.YMSSP.2015.10.013.

[20] X. P. Yan, C. H. Zhao, Z. Y. Lu, X. C. Zhou, and H. L. Xiao, "A study of information technology used in oil monitoring," *Tribology International*, vol. 38, no. 10, pp. 879–886, Oct. 2005, doi: 10.1016/J.TRIBOINT.2005.03.012.

[21] H. Towsyfyan, F. Gu, A. D. Ball, and B. Liang, "Tribological behaviour diagnostic and fault detection of mechanical seals based on acoustic emission measurements," *Friction*, vol. 7, no. 6, pp. 572–586, Nov. 2018, doi: 10.1007/S4 0544-018-0244-4.

[22] J. Moder, P. Bergmann, and F. Grün, "Lubrication regime classification of hydrodynamic journal bearings by machine learning using torque data," *Lubricants*, vol. 6, no. 4, p. 108, Dec. 2018, doi: 10.3390/LUBRICANTS604 0108.

[23] C. Chimeno-Trinchet, C. Murru, M. E. Díaz-García, A. Fernández-González, and R. Badía-Laíño, "Artificial intelligence and fourier-transform infrared spectroscopy for evaluating water-mediated degradation of lubricant oils," *Talanta*, vol. 219, p. 121312, Nov. 2020, doi: 10.1016/J.TALANTA.202 0.121312.

[24] B. Peng, T. Bergs, D. Schraknepper, F. Klocke, and B. Döbbeler, "A hybrid approach using machine learning to predict the cutting forces under consideration of the tool wear," *Procedia CIRP*, vol. 82, pp. 302–307, Jan. 2019, doi: 10.1016/ J.PROCIR.2019.04.031.

[25] F. E. Bock, L. A. Blaga, and B. Klusemann, "Mechanical performance prediction for friction riveting joints of dissimilar materials via machine learning," *Procedia Manufacturing*, vol. 47, pp. 615–622, Jan. 2020, doi: 10.1016/ J.PROMFG.2020.04.189.

[26] H. Zhang *et al.*, "A generalized artificial intelligence model for estimating the friction angle of clays in evaluating slope stability using a deep neural network

and Harris Hawks optimization algorithm," *Engineering with Computers*, pp. 1–14, Jan. 2021, doi: 10.1007/S00366-020-01272-9.

[27] Z. Tariq, S. Elkatatny, M. Mahmoud, A. Z. Ali, and A. Abdulraheem, "A new approach to predict failure parameters of carbonate rocks using artificial intelligence tools," *SPE Kingdom of Saudi Arabia Annual Technical Symposium and Exhibition*, pp. 1428–1440, Apr. 2017, doi: 10.2118/187974-MS.

[28] Z. Luo, X.-N. Bui, H. Nguyen, and H. Moayedi, "A novel artificial intelligence technique for analyzing slope stability using PSO-CA model," *Engineering with Computers*, vol. 37, no. 1, pp. 533–544, Aug. 2019, doi: 10.1007/S00366-019-00839-5.

[29] Y. Zeng, D. Song, W. Zhang, B. Zhou, M. Xie, and X. Tang, "A new physics-based data-driven guideline for wear modelling and prediction of train wheels," *Wear*, vol. 456–457, p. 203355, Sep. 2020, doi: 10.1016/J.WEAR.2020.203355.

[30] T. Thankachan, K. Soorya Prakash, V. Kavimani, and S. R. Silambarasan, "Machine learning and statistical approach to predict and analyze wear rates in copper surface composites," *Metals and Materials International*, vol. 27, no. 2, pp. 220–234, Jul. 2020, doi: 10.1007/S12540-020-00809-3.

[31] A. Bustillo, D. Y. Pimenov, M. Matuszewski, and T. Mikolajczyk, "Using artificial intelligence models for the prediction of surface wear based on surface isotropy levels," *Robotics and Computer-Integrated Manufacturing*, vol. 53, pp. 215–227, Oct. 2018, doi: 10.1016/J.RCIM.2018.03.011.

[32] A. Gouarir, G. Martínez-Arellano, G. Terrazas, P. Benardos, and S. Ratchev, "In-process tool wear prediction system based on machine learning techniques and force analysis," *Procedia CIRP*, vol. 77, pp. 501–504, Jan. 2018, doi: 10.1016/J.PROCIR.2018.08.253.

[33] T. Thankachan, K. Soorya Prakash, and M. Kamarthin, "Optimizing the tribological behavior of hybrid copper surface composites using statistical and machine learning techniques," *Journal of Tribology*, vol. 140, no. 3, p. 031610, May 2018, doi: 10.1115/1.4038688.

[34] J. Wang, Y. Li, R. Zhao, and R. X. Gao, "Physics guided neural network for machining tool wear prediction," *Journal of Manufacturing Systems*, vol. 57, pp. 298–310, Oct. 2020, doi: 10.1016/J.JMSY.2020.09.005.

[35] S. Shankar, T. Mohanraj, and R. Rajasekar, "Prediction of cutting tool wear during milling process using artificial intelligence techniques," *International Journal of Computer Integrated Manufacturing*, vol. 32, no. 2, pp. 174–182, Feb. 2018, doi: 10.1080/0951192X.2018.1550681.

[36] J. Karandikar, "Machine learning classification for tool life modeling using production shop-floor tool wear data," *Procedia Manufacturing*, vol. 34, pp. 446–454, Jan. 2019, doi: 10.1016/J.PROMFG.2019.06.192.

[37] O. Altay, T. Gurgenc, M. Ulas, and C. Özel, "Prediction of wear loss quantities of ferro-alloy coating using different machine learning algorithms," *Friction*, vol. 8, no. 1, pp. 107–114, Jan. 2019, doi: 10.1007/S40544-018-0249-Z.

[38] S. Bitrus, I. Velkavrh, and E. Rigger, "Applying an adapted data mining methodology (DMME) to a tribological optimisation problem," in *Data Science – Analytics and Applications*. Springer, 2021, pp. 38–43, doi: 10.1007/978-3-658-321 82-67.

[39] A. de Farias, S. L. R. de Almeida, S. Delijaicov, V. Seriacopi, and E. C. Bordinassi, "Simple machine learning allied with data-driven methods for monitoring tool wear in machining processes," *The International Journal of Advanced Manufacturing Technology*, vol. 109, no. 9, pp. 2491–2501, Aug. 2020, doi: 10.1007/S00170-020-05785-X.

[40] T. Kolodziejczyk, R. Toscano, S. Fouvry, and G. Morales-Espejel, "Artificial intelligence as efficient technique for ball bearing fretting wear damage prediction," *Wear*, vol. 268, no. 1–2, pp. 309–315, Jan. 2010, doi: 10.1016/J.WEAR.2009.08.016.

[41] M. Bellotti, M. Wu, J. Qian, and D. Reynaerts, "Tool wear and material removal predictions in micro-EDM drilling: Advantages of data-driven approaches," *Applied Sciences*, vol. 10, no. 18, p. 6357, Sep. 2020, doi: 10.3390/APP10186357.

[42] G. Kronberger *et al.*, "Using robust generalized fuzzy modeling and enhanced symbolic regression to model tribological systems," *Applied Soft Computing*, vol. 69, pp. 610–624, Aug. 2018, doi: 10.1016/J.ASOC.2018.04.048.

[43] A. Tran, J. M. Furlan, K. V. Pagalthivarthi, R. J. Visintainer, T. Wildey, and Y. Wang, "WearGP: A computationally efficient machine learning framework for local erosive wear predictions via nodal Gaussian processes," *Wear*, vol. 422–423, pp. 9–26, Mar. 2019, doi: 10.1016/J.WEAR.2018.12.081.

[44] M. Hayajneh, A. M. Hassan, A. Alrashdan, and A. T. Mayyas, "Prediction of tribological behavior of aluminum–copper based composite using artificial neural network," *Journal of Alloys and Compounds*, vol. 470, no. 1–2, pp. 584–588, Feb. 2009, doi: 10.1016/J.JALLCOM.2008.03.035.

[45] F. Aydin, "The investigation of the effect of particle size on wear performance of AA7075/Al$_2$O$_3$ composites using statistical analysis and different machine learning methods," *Advanced Powder Technology*, vol. 32, no. 2, pp. 445–463, Feb. 2021, doi: 10.1016/J.APT.2020.12.024.

[46] A. Borjali, K. Monson, and B. Raeymaekers, "Predicting the polyethylene wear rate in pin-on-disc experiments in the context of prosthetic hip implants: Deriving a data-driven model using machine learning methods," *Tribology International*, vol. 133, pp. 101–110, May 2019, doi: 10.1016/J.TRIBOINT.201 9.01.014.

[47] X. Gao, K. Dai, Z. Wang, T. Wang, and J. He, "Establishing quantitative structure tribo-ability relationship model using Bayesian regularization neural network," *Friction*, vol. 4, no. 2, pp. 105–115, Mar. 2016, doi: 10.1007/S40544-016-0104-Z.

[48] R. Egala, G. V. Jagadeesh, and S. G. Setti, "Experimental investigation and prediction of tribological behavior of unidirectional short castor oil fiber reinforced epoxy composites," *Friction*, vol. 9, no. 2, pp. 250–272, Jul. 2020, doi: 10.1007/ S40544-019-0332-0.

[49] A. Kordijazi *et al.*, "Data-driven modeling of wetting angle and corrosion resistance of hypereutectic cast Aluminum-Silicon alloys based on physical and chemical properties of surface," *Surfaces and Interfaces*, vol. 20, p. 100549, Sep. 2020, doi: 10.1016/J.SURFIN.2020.100549.

[50] F. Alambeigi, S. M. Khadem, H. Khorsand, and E. Mirza Seied Hasan, "A comparison of performance of artificial intelligence methods in prediction of dry sliding wear behavior," *The International Journal of Advanced Manufacturing Technology*, vol. 84, no. 9, pp. 1981–1994, Sep. 2015, doi: 10.1007/S00170-015-7812-9.

[51] X. Xu *et al.*, "Machine learning-based wear fault diagnosis for marine diesel engine by fusing multiple data-driven models," *Knowledge-Based Systems*, vol. 190, p. 105324, Feb. 2020, doi: 10.1016/J.KNOSYS.2019.105324.

[52] M. Cheng, L. Jiao, X. Shi, X. Wang, P. Yan, and Y. Li, "An intelligent prediction model of the tool wear based on machine learning in turning high strength steel," *Proceedings of the Institution of Mechanical Engineers, Part B:*

Journal of Engineering Manufacture, vol. 234, no. 13, pp. 1580–1597, Jul. 2020, doi: 10.1177/0954405420935787.

[53] K. Gouda, P. Rycerz, A. Kadiric, and G. Morales-Espejel, "Assessing the effectiveness of data-driven time-domain condition indicators in predicting the progression of surface distress under rolling contact," *Proceedings of the Institution of Mechanical Engineers, Part J: Journal of Engineering Tribology*, vol. 233, no. 10, pp. 1523–1540, Mar. 2019, doi: 10.1177/1350650119838896.

[54] H. Moayedi and S. Hayati, "Artificial intelligence design charts for predicting friction capacity of driven pile in clay," *Neural Computing and Applications*, vol. 31, no. 11, pp. 7429–7445, Jun. 2018, doi: 10.1007/S00521-018-3555-5.

[55] S. Tatipala, J. Wall, C. M. Johansson, and M. Sigvant, "Data-driven modelling in the era of Industry 4.0: A case study of friction modelling in sheet metal forming simulations," *Journal of Physics: Conference Series*, vol. 1063, no. 1, p. 012135, Jul. 2018, doi: 10.1088/1742-6596/1063/1/012135.

[56] G. Boidi, M. R. da Silva, F. J. Profito, and I. F. Machado, "Using machine learning radial basis function (RBF) method for predicting lubricated friction on textured and porous surfaces," *Surface Topography: Metrology and Properties*, vol. 8, no. 4, p. 044002, Nov. 2020, doi: 10.1088/2051-672X/ABAE13.

[57] N. Parveen, S. Zaidi, and M. Danish, "Artificial intelligence (AI)-based friction factor models for large piping networks," *Chemical Engineering Communications*, vol. 207, no. 2, pp. 213–230, Feb. 2019, doi: 10.1080/00986445.2019.1578757.

[58] M. Najafzadeh, J. Shiri, G. Sadeghi, and A. Ghaemi, "Prediction of the friction factor in pipes using model tree," *ISH Journal of Hydraulic Engineering*, vol. 24, no. 1, pp. 9–15, Jan. 2017, doi: 10.1080/09715010.2017.1333926.

[59] M. Niazkar, "Revisiting the estimation of colebrook friction factor: A comparison between artificial intelligence models and C-W based explicit equations," *KSCE Journal of Civil Engineering*, vol. 23, no. 10, pp. 4311–4326, Sep. 2019, doi: 10.1007/S12205-019-2217-1.

[60] D. Gropper, L. Wang, and T. J. Harvey, "Hydrodynamic lubrication of textured surfaces: A review of modeling techniques and key findings," *Tribology International*, vol. 94, pp. 509–529, Feb. 2016, doi: 10.1016/j.triboint.2015.10.009.

[61] A. Kurdi, N. Alhazmi, H. Alhazmi, and T. Tabbakh, "Practice of simulation and life cycle assessment in tribology—A review," *Materials*, vol. 13, no. 16, p. 3489, Aug. 2020, doi: 10.3390/MA13163489.

[62] Y. Zhang and X. Xu, "Machine learning decomposition onset temperature of lubricant additives," *Journal of Materials Engineering and Performance*, vol. 29, no. 10, pp. 6605–6616, Oct. 2020, doi: 10.1007/S11665-020-05146-5.

[63] P. S. Desai, V. Granja, and C. F. Higgs, "Lifetime prediction using a tribology-aware, deep learning-based digital twin of ball bearing-like tribosystems in oil and gas," *Processes*, vol. 9, no. 6, p. 922, May 2021, doi: 10.3390/PR9060922.

[64] V. Zambrano *et al.*, "A digital twin for friction prediction in dynamic rubber applications with surface textures," *Lubricants*, vol. 9, no. 5, p. 57, May 2021, doi: 10.3390/LUBRICANTS9050057.

[65] A. Díaz Lantada, F. Franco-Martínez, S. Hengsbach, F. Rupp, R. Thelen, and K. Bade, "Artificial intelligence aided design of microtextured surfaces: Application to controlling wettability," *Nanomaterials*, vol. 10, no. 11, pp. 1–19, Nov. 2020, doi: 10.3390/NANO10112287.

[66] A. Kordijazi, S. Behera, D. Patel, P. Rohatgi, and M. Nosonovsky, "Predictive analysis of wettability of Al–Si based multiphase alloys and aluminum matrix composites by machine learning and physical modeling," *Langmuir*, vol. 37, no. 12, pp. 3766–3777, Mar. 2021, doi: 10.1021/ACS.LANGMUIR.1C00358.

[67] L. Cavaleri, P. G. Asteris, P. P. Psyllaki, M. G. Douvika, A. D. Skentou, and N. M. Vaxevanidis, "Prediction of surface treatment effects on the tribological performance of tool steels using artificial neural networks," *Applied Sciences*, vol. 9, no. 14, p. 2788, Jul. 2019, doi: 10.3390/APP9142788.

[68] R. M KordijaziAmir, A. Dhingra, R. K PovoloMarco, and M. Nosonovsky, "Machine-learning methods to predict the wetting properties of iron-based composites," *Surface Innovations*, vol. 9, no. 2–3, pp. 111–119, Jan. 2021, doi: 10.1680/JSUIN.20.00024.

[69] M. O. Shabani and A. Mazahery, "Artificial intelligence in numerical modeling of nano sized ceramic particulates reinforced metal matrix composites," *Applied Mathematical Modelling*, vol. 36, no. 11, pp. 5455–5465, Nov. 2012, doi: 10.1016/J.APM.2011.12.059.

[70] T. Banerjee, S. Dey, A. P. Sekhar, S. Datta, and D. Das, "Design of alumina reinforced aluminium alloy composites with improved tribo-mechanical properties: A machine learning approach," *Transactions of the Indian Institute of Metals*, vol. 73, no. 12, pp. 3059–3069, Oct. 2020, doi: 10.1007/S12666-020-02108-2.

[71] A. Vinoth and S. Datta, "Design of the ultrahigh molecular weight polyethylene composites with multiple nanoparticles: An artificial intelligence approach," *Journal of Composite Materials*, vol. 54, no. 2, pp. 179–192, Jul. 2019, doi: 10.1177/0021998319859924.

[72] T. Sathish, "BONN technique: Tribological properties predictor for plasma nitrided 316L stainless steel," *Materials Today: Proceedings*, vol. 5, no. 6, pp. 14545–14552, Jan. 2018, doi: 10.1016/J.MATPR.2018.03.044.

[73] E. W. Bucholz *et al.*, "Data-driven model for estimation of friction coefficient via informatics methods," *Tribology Letters*, vol. 47, no. 2, pp. 211–221, May 2012, doi: 10.1007/S11249-012-9975-Y.

[74] D. Wilfred, T. Tysoe, and N. D. Spencer, "Designing lubricants by artificial intelligence," *Tribology & Lubrication Technology*, vol. 76, no. 6, p. 78, 2020.

[75] S. Bhaumik, S. D. Pathak, S. Dey, and S. Datta, "Artificial intelligence based design of multiple friction modifiers dispersed castor oil and evaluating its tribological properties," *Tribology International*, vol. 140, p. 105813, Dec. 2019, doi: 10.1016/J.TRIBOINT.2019.06.006.

[76] S. Bhaumik and M. Kamaraj, "Artificial neural network and multi-criterion decision making approach of designing a blend of biodegradable lubricants and investigating its tribological properties," *Proceedings of the Institution of Mechanical Engineers, Part J: Journal of Engineering Tribology*, vol. 235, no. 8, pp. 1575–1589, Oct. 2020, doi: 10.1177/1350650120965754.

[77] C. Humelnicu, S. Ciortan, and V. Amortila, "Artificial neural network-based analysis of the tribological behavior of vegetable oil–diesel fuel mixtures," *Lubricants*, vol. 7, no. 4, p. 32, Apr. 2019, doi: 10.3390/LUBRICANTS7040032.

[78] T. Ramkumar and M. Selvakumar, *Artificial Intelligence in Predicting the Optimized Wear Behaviour Parameters of Sintered Titanium Grade 5 Reinforced with Nano B4C Particles*. CRC Press, 2021, doi: 10.1201/9781003011248-2.

[79] A. Babin, "Data-driven system identification and optimal control of an active rotor-bearing system," *IOP Conference Series: Materials Science and*

Engineering, vol. 1047, no. 1, p. 012053, Feb. 2021, doi: 10.1088/1757-899X/1 047/1/012053.

[80] K. Wierzcholski and O. Łupicka, "Network analysis for artificial intelligent micro-bearing systems," *Journal of KONES*, vol. 16, no. 3, pp. 489–495, 2009.

[81] J. Y. Li, X. X. Yao, and Z. Zhang, "Physical model based on data-driven analysis of chemical composition effects of friction stir welding," *Journal of Materials Engineering and Performance*, vol. 29, no. 10, pp. 6591–6604, Sep. 2020, doi: 10.1007/S11665-020-05132-X.

4 Case studies

4.1 TOOL WEAR PREDICTION

Tool wear has a critical impact on the quality of workpieces. Excessive wear can lead to a decrease in processing accuracy and speed and a yield decline. To avoid excessive tool wear, frequent tool changes might happen, but frequent tool changes increase costs and have a negative impact on processing speed. Therefore, it is essential to recognize tool wear status more accurately to plan a more reasonable tool change time and optimize the tool design.

It has been proved that data-driven wear prediction based on multi-sensor signals processing is a feasible solution to tool wear identification. However, the model that is trained in a specific working condition does not fit another condition in most cases. The tool wear characteristics can be different even if an identical tool is put under the same working conditions. This limits the actual application of data-driven wear prediction methods in industries. In most cases, the acquisition of tool wear data is achieved through offline detection, which means the tool is observed with microscopes after it is taken off processing equipment. This results in low efficiency of data labeling.

Tribo-informatics can play a big role in solving this problem. Several steps are needed:

1. The first one is to conduct experiments on the wear prediction of face milling cutters. Engineers should select force/torque sensors, vibration sensors, and acoustic emission sensors to record signals that are generated during the processing of workpieces. These signals can serve as input sources for training and prediction models. In addition, to get the ground truth of the wear, a CMOS camera should be installed on the processing platform to capture pictures of the cutter wear position.

2. The second step is to extract and filter original signals in order to reduce interference of those non-cutting factors. Multiple statistics for every signal from the time and frequency domain are extracted, serving as signal features. This research also proposes a machine vision-based method of tool wear recognition. With prerequisite knowledge of wear morphologies of the face milling cutters, this method uses an image preprocessing method. This sample research uses canny edge extraction and region recognition in order to achieve in-situ tool wear condition acquisition and to avoid frequent disassembly of cutters.

3. The third step is research on wear prediction algorithms in single working condition scenarios. Using the above signal features and wear features, a

DOI: 10.1201/9781003467991-4

tool wear prediction model is established in a single operating scenario using methods such as support vector machine, random forest, and backpropagation neural network. The optimal hyperparameters of each model are obtained. The training time, prediction time, and result accuracy of different algorithms are compared. The results show that the accuracy (determination coefficient) of these three algorithms in the test set under single operating conditions is above 0.99, verifying the feasibility of the multi-source information prediction wear method.

4. The fourth one is the application of transfer learning method in multi-working condition tool wear scenarios. A wear prediction method based on feature transfer is proposed for tool wear prediction under multiple working conditions. By comparing historical features with the previous data of the current tool, using root mean square difference as the loss function, a one-to-one linear transfer model is solved, and the maximum mean difference is used as the evaluation function value to filter out successful transfer features. These features are used to establish a migrated wear prediction model, achieving optimized prediction of new working conditions. The application results show that the wear prediction model established using source data has bad prediction performance under two target working conditions. However, the model using pre target working condition data for feature transfer can significantly improve prediction accuracy, and the accuracy evaluation index R2 determination coefficient is increased from about 0.4 to 0.96, verifying the effectiveness of the wear prediction method proposed in this sample research based on transfer learning and multi-source information fusion.

This section will first introduce the research background, including the application value of tool wear monitoring and the research value of the tool wear prediction method based on multi-source information fusion. Then, this section will cover surface milling cutter wear experiment, visual wear features and signal feature extraction, and a method for predicting tool wear under multiple working conditions based on transfer learning.

4.1.1 RESEARCH BACKGROUND OF TOOL WEAR PREDICTION

With the development of industrial modernization, the demand for machining accuracy is also constantly increasing. As an important component of using machine tools, tool life prediction and wear status monitoring are crucial. Reasonable determination of tool change time can significantly improve industrial production efficiency and product quality. This tool wear prediction project aims to integrate multiple sensor information during machine tool processing, monitor the wear status of cutting tools, and predict their remaining service life. The background and significance of the project are mainly divided into two parts: the application value of tool wear status monitoring and the research value of multi-source information fusion tool wear prediction methods.

4.1.1.1   Application value of tool wear status monitoring

Tool wear is mainly caused by the friction between the tool and the workpiece, as well as between the tool and chips during the machining process. According to the national standard GB/T 16460-2016 "End milling cutter life test," the tool life is defined as the total cutting time of the tool. Due to different cutting conditions, there are various tool failure phenomena. Among them, the most commonly used tool wear judgment is based on the wear width of the rear tool surface. For uniform wear, an average wear width of 0.3 mm can be used as the condition for the end of tool service life. For local wear, the maximum wear width of 0.5 mm on any tooth can be used as a condition for the end of tool life. In addition, the depth value of previous wear can also serve as a reference for lifespan judgment.

The increase in tool wear with increasing cutting times is not linear. In fact, the overall wear curve can be divided into three stages: initial rapid wear period I, stable wear period II, and rapid wear period III. The moment when the tool reaches a sharp wear state is usually considered as the time to replace the tool.

The quality of the product is closely related to the machining process, which mainly depends on the state of the cutting tool. Excessive wear and tear on cutting tools can lead to a decrease in machining accuracy and speed, end products with substandard quality, waste of raw materials and time, and even machine tool damage, seriously affecting the work process. On the other hand, premature tool replacement cannot fully utilize the service life of the tool, resulting in an increase in tool costs. Considering safety and operability reasons, changing the tool requires stopping the machine's operation. Research has shown that 20% of downtime is caused by tool failures, and frequent downtime and tool changes greatly affect work and production progress. Accurately determining the time for tool replacement can increase the processing speed by 10–50%, achieving the goal of fully utilizing the tool's service life while ensuring processing accuracy and high-speed operation of the production line.

At present, many enterprises still rely on technical personnel to determine whether tool replacement is necessary based on the processed products. This method has a certain degree of feasibility, but it is only applicable to situations where the price of processed parts is not high and the single processing process is not long. For materials with high value and difficulty in processing, such as some aviation parts, the tool needs to be replaced multiple times during the processing, so it cannot be judged after the processing is completed. At present, the research on the surface morphology of processed products lacks the timeliness of tool wear status and tool change time judgment during the processing. It is necessary to develop a real-time tool status in-place monitoring system.

4.1.1.2   Research value of tool wear prediction method

Tool wear prediction based on sensor information can generally be divided into two methods. One is based on physical theoretical models, and the other is based on data and artificial intelligence algorithm models. The prediction based on physical theoretical models requires the establishment of a calculation model between wear amount and machining parameters as well as environmental variables based on experimental and simulation results. This method has strong scientific accuracy. However, the establishment of real models is extremely complex, and the forms of tool wear are diverse and greatly affected by accidental factors. This results in a large error in predicting the wear amount of the simplified model. Therefore, the use of data-driven tool for life prediction is more common and efficient.

With the vigorous development of sensor technology, mass storage technology, and high-performance computing technology, the application of artificial intelligence in the manufacturing industry has become an inevitable trend. In recent years, many advanced intelligent manufacturing technologies have emerged, and the manufacturing industry has also moved from the traditional mechanical age to the information and big data age.

A large amount of sensor data and processing parameter information is generated during the manufacturing process. Reasonable collection and in-depth analysis of these data can help better grasp the production process, monitor equipment status, identify and handle potential problems, improve weak parts of the production process, and improve production efficiency. In addition, single signal monitoring may be affected by environmental factors and have an impact on the results. The prediction method of multi-source information fusion can supplement various information, improve the stability and accuracy of tool wear prediction models, and thus has high research value.

4.1.2 SURFACE MILLING CUTTER WEAR EXPERIMENT

4.1.2.1 Introduction

Face milling cutters are widely used in flat machining of hard alloys due to their replaceable blades, long service life, and low average processing costs. However, the existing tool wear experiments mostly focus on turning tools and end mills, while the wear exploration experiments on surface milling cutters are insufficient.

Therefore, this section conducts wear experiments on coated surface milling cutters to explore their wear forms and patterns. Firstly, considering the experimental equipment conditions comprehensively, this section designs the experimental plan and builds the experimental platform. Subsequently, the experimental steps are clarified, and based on the obtained experimental results, the wear patterns of the coated surface milling cutter are analyzed and compared with those of the end milling cutter.

4.1.2.2 Experimental equipment and software

The experimental equipment and software are shown in Figure 4.1a, which shows the Deckel Maho DMU50 five-axis milling CNC machine tool. Speed range: 20–18000 rpm. Positioning accuracy: 0.008 μm. X/Y/Z axis movement range: 500/450/400 mm. Maximum movement speed: 24 m/min. Tool handle model: ISO40 DIN 69871. Tool table range: 630 × 500 mm.

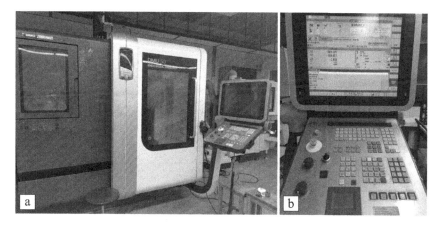

FIGURE 4.1   Deckel Maho DMU50 Five-axis milling CNC machine. a) Machine; b) CNC panel.

The CNC panel of this machine tool is shown in Figure 4.1b, which supports milling trajectory programming with global variables, and the machining process can run in steps, making it easy for sensors to record data in a timely manner.

The image data are collected using the Beijing Airlines JHSM1400f color industrial camera (Figure 4.2a). The highest pixel is 14 million, and imaging parameters such as brightness, exposure, and tone can be adjusted. Figure 4.2b shows the corresponding software interface, which supports setting the region of interest (ROI) to obtain high-resolution blade images and significantly reduce the memory required for image storage. In addition, the camera software also supports adding markings in the imaging window, making it easy to confirm the consistency of each imaging position.

The collection of force and torque during the machining process is carried out using the Kistler 9257B contact triaxial force measuring instrument (Figure 4.3a). This force measuring instrument can measure forces and moments in three directions

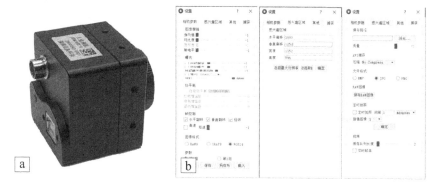

FIGURE 4.2   JHSM1400f CMOS camera. a) Camera; b) Software.

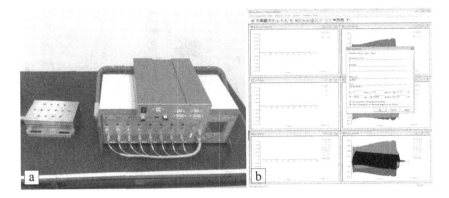

FIGURE 4.3 Kistler 5697A three-axis dynamometer. a) Dynamometer platform, signal amplifier, acquisition card; b) Software.

X/Y/Z, with a range of ± 1000 N and an accuracy of 0.01 N. The surface size of the dynamometer is 100 × 170 mm, supporting direct fixation of the workpiece on the dynamometer. The force and torque signals detected by the dynamometer need to be converted into voltage signals through a charge amplifier and acquisition card and transmitted to a computer. The software DynoWare (Figure 4.3b) can be used to view the signals in real time and export them to CSV format.

The vibration sensor is similar to the acoustic emission sensor (Figure 4.4a), with a strong magnet at the head that can be directly adsorbed on the magnetic

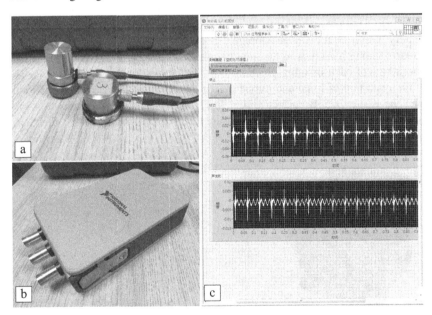

FIGURE 4.4 Vibration sensor and acoustic emission sensor. a) Sensor; b) NI Capture Card; c) Labview.

workpiece for easy installation. The output data is a voltage signal, which can be observed and stored in real time using the software Labview (Figure 4.4c) after passing through the acquisition card (Figure 4.4b).

The HR-150DT metal Rockwell hardness tester is used to measure the hardness of workpieces. This hardness tester can automatically add and remove test forces, with simple operation, high sensitivity, and stability, and it is suitable for measuring the hardness of metals such as steel and copper.

4.1.2.3 Design of experimental plan

4.1.2.3.1 Construction of experimental platform

The surface milling cutter cutting experiment is conducted in a five-axis CNC machine tool. The experimental platform is built as shown in Figure 4.5. The industrial camera is fixed with a vise on the left side of the machine tool, and a 4 mm lens is used to capture wear images in place. The main body of the vibration and acoustic emission sensor is directly installed on the workpiece through magnetic attraction, and the workpiece is fixed on the triaxial dynamometer. The collected sensor signal data are converted into voltage signals by the acquisition card and transmitted to the computer for subsequent analysis and research. The *X/Y/Z* direction has been marked in the figure, where *X* is the tool feed direction and *Z* is the tool axis.

4.1.2.3.2 Material and processing parameter selection

The face milling cutter uses a mountain high four-blade cutter bar with a cutting diameter of 50 mm, and the blade model is SEMX1204AFTN-M15 F40M, suitable for processing harder materials such as steel. During the machining

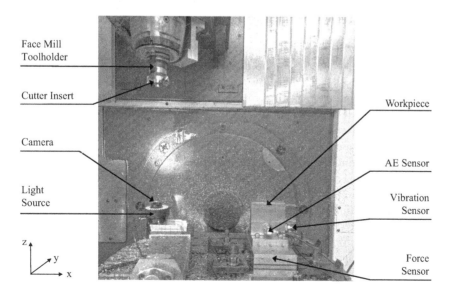

FIGURE 4.5 Experimental setup.

TABLE 4.1

Parameters of Processing

No	v_f (mm/min)	n (r/min)	a_p (mm)	f (mm/r)	material	h (HRC)
1	80	800	0.9	0.1	1	27.3
2	60	1200	1.2	0.05	1	27.4
3	80	1000	1.2	0.08	1	27.4
4	100	1000	0.6	0.1	2	9.7
5	70	1000	1.5	0.07	2	10.1
6	100	1000	0.6	0.1	3	5.4
7	60	800	0.6	0.075	3	5.3

process, the fixed parameters are 50% cutting width and a sampling frequency of 4k Hz. The adjustable machining parameters are feed rate (vf), rotational speed (n), and cutting depth (a_p), and the feed rate per revolution (f) can be calculated. This experiment selected three types of steel, two types of stainless steel, and one type of carbon steel, as the materials for processing workpieces. Due to the significant influence of material hardness on the friction coefficient, the average hardness (h) of 10 sampling points on the surface of the workpiece is taken as the material parameter.

A total of six sets of surface milling cutter wear experiments were conducted. The first, second, and third groups of experiments used the No. 1 stainless steel workpiece with the highest hardness, the fourth and fifth groups of experiments used the No. 2 stainless steel workpiece with medium hardness, and the sixth and seventh groups of experiments used the carbon steel workpiece with the lowest hardness. The specific processing parameters of the experiment are shown in Table 4.1.

4.1.2.3.3 Experimental procedures

The experimental platform can be divided into three areas according to their functions, namely the photography area, processing area, and collection area (Figure 4.6). Before the experiment begins, move the milling cutter to the photo area and record the image of the blade when it is not worn. Then, start the cutting cycle, cutting off one layer of workpiece material each time. During the cutting process, sensors are used to collect a total of eight signals, including force and torque signals from three directions, vibration signals in the x-direction, and acoustic emission signals in the y-direction. After the single-layer cutting is completed, it returns to the photography area again and records the wear image after the completion of the cutting. At this time, a cutting cycle is completed. Each experiment starts with the blade not worn and stops when the tool is excessively worn or the workpiece is completely consumed.

The specific process of the experiment is shown in Figure 4.7. The experimental output includes images of tool wear, as well as force signals, torque signals, vibration signals, acoustic emission signals, and tool wear images corresponding to each cutting.

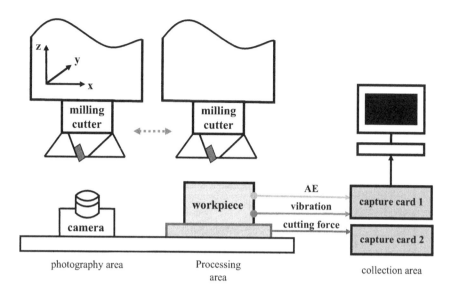

photography area   Processing   collection area
                                   area

FIGURE 4.6 Simplified diagram of the processing platform.

4.1.2.4 Preliminary analysis of experimental results

4.1.2.4.1 Tool wear image

The tool image can intuitively reflect the wear of the tool. Through image analysis, not only can the wear amount be detected, but also the relationship between the wear form and working conditions can be analyzed. In each group of experiments, images of the tool without wear and 50 images of the tool at the completion of processing are recorded. Figure 4.8 extracts and displays the tool wear images in each group of experiments. The experimental results show that the form of tool wear is most affected by the hardness of the workpiece material, and there are two types of wear for face milling cutters: coating detachment and pits, especially for machining harder materials.

Figure 4.8a shows the image of the tool when it is not worn, from which the complete edge of the tool can be obtained, which can then be compared with the worn tool image for wear detection.

Figure 4.8b, c, and d shows the tool wear images after the 40th cutting in Experiments 1, 2, and 3, respectively. The workpiece material used is stainless steel with the highest hardness. From the figure, it can be seen that both the auxiliary and rear cutting surfaces are worn. The wear form of the auxiliary rear blade surface is coating detachment, and the wear amount is relatively small. There are two forms of wear on the rear blade surface: pits and coating detachment, which have a significant amount of wear.

Figure 4.8e and f shows the tool wear images after the 40th cutting in Experiments 4 and 5, respectively. The workpiece material used is stainless steel with lower hardness. It can be seen that after the same amount of processing, the

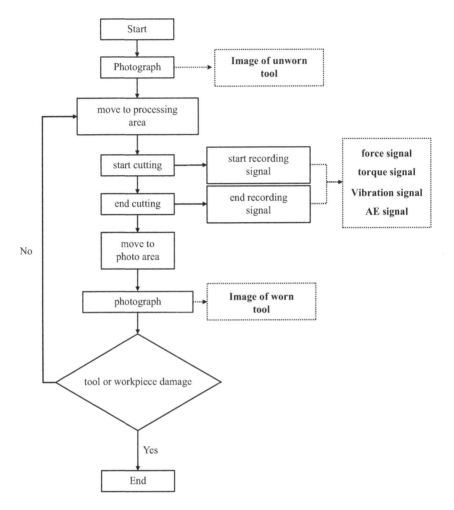

FIGURE 4.7   Flow chart of experiment.

overall wear of the tool is relatively small, and there are no pits. In addition, the length of the worn edge of the rear cutting surface is positively correlated with the cutting depth.

Figure 4.8g and h shows the tool wear images after the 40th cutting in Experiments 6 and 7, respectively. The workpiece material used is No. 45 carbon steel with the lowest hardness. It can be seen that there is a uniform wear zone on the back face of the auxiliary blade, and its wear width is significantly greater than that of other materials in the experiment. In addition, in these two sets of experiments, the wear of the rear blade surface was relatively small.

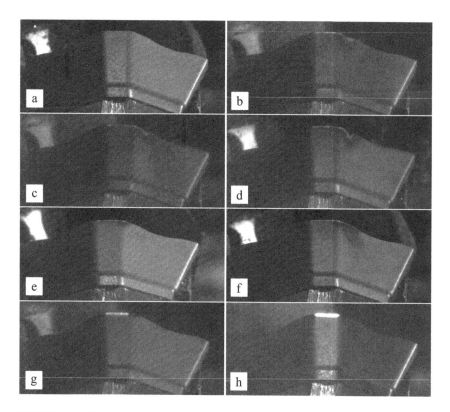

FIGURE 4.8 Photo of new tool and photo of tool wear after 40th cutting in each experiment. a) Not worn; b) Experiment 1; c) Experiment 2; d) Experiment 3; e) Experiment 4; f) Experiment 5; g) Experiment 6; h) Experiment 7.

4.1.2.4.2 Cutting force and torque signals

We extract the force signals corresponding to a single tool run (i.e., tool rotation for one cycle) from the 40th set of cutting signal data in different experiments, as shown in Figure 4.9. For a single feed, the cutting force in the X-direction (feed direction) increases sharply during feed, then slowly decreases, while the cutting force in the Y-direction is the opposite, first slowly increasing and then sharply decreasing. The Z-direction (axial) cutting force is relatively stable throughout the entire cutting process.

Overall, the signal shapes of different experiments are similar and have transferability. The material hardness has a significant impact on the amplitude of force, and the force signal amplitudes corresponding to Experiments 1, 2, and 3 of processing the stainless steel workpiece with the highest hardness are much greater than those corresponding to other experiments. For the same workpiece material, changes in operating parameters also have a certain impact on the shape and amplitude of the signal.

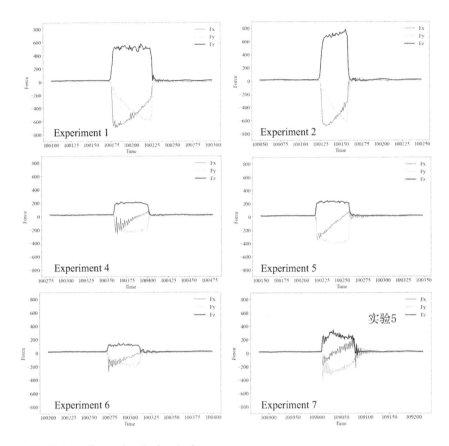

FIGURE 4.9 Force signal of a single cut.

In addition, all force signals have a certain level of noise, among which the workpiece materials used in Experiments 1, 2, and 3 are harder, resulting in a larger amplitude of cutting force. Therefore, the noise has a relatively small impact and does not affect the overall waveform. In order to reduce the interference of noise on subsequent wear prediction results, the original signal can be filtered and processed.

By extracting the same single cutting torque signal (Figure 4.10), it can be seen that the torque signal is similar to the force signal under different working conditions and materials. For harder materials, the overall amplitude of the signal is larger, and the impact of noise is relatively small.

4.1.2.4.3 Vibration signals and acoustic emission signals

We extract the vibration and acoustic emission signals of the 40th cutting during a single cutting in each group of experiments, as shown in Figure 4.11. From the signal of the cutting gap, it can be seen that there are low-frequency and low-amplitude environmental signals that can be eliminated through filtering. In

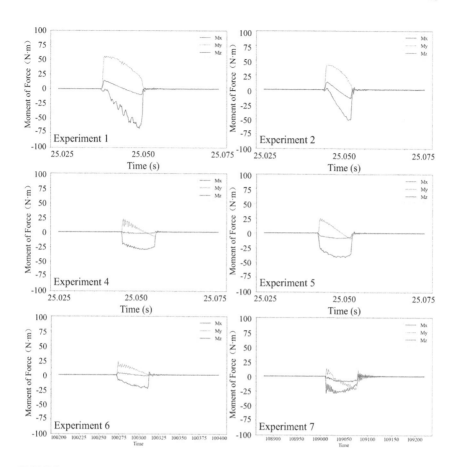

FIGURE 4.10 Moment signal of a single cut.

addition, due to the high frequency of signals generated by cutting, they are more severely affected by high-frequency noise interference, resulting in significant fluctuations in signal amplitude over time.

Compared to cutting force and torque signals, vibration and acoustic emission signals have significant waveform differences under different working conditions, and their regularity is difficult to obtain through signal curves. Therefore, feature extraction and analysis are needed. However, it can still be seen that the amplitude of the vibration signal and acoustic emission signal increases overall with the decrease of the workpiece hardness, with the increase in the amplitude of the acoustic emission signal being more significant.

4.1.2.5 Summary
This section conducts wear experiments on coated surface milling cutters. First, the experimental design includes an introduction to the experimental equipment and parameters, the construction of the experimental platform, the selection of workpiece

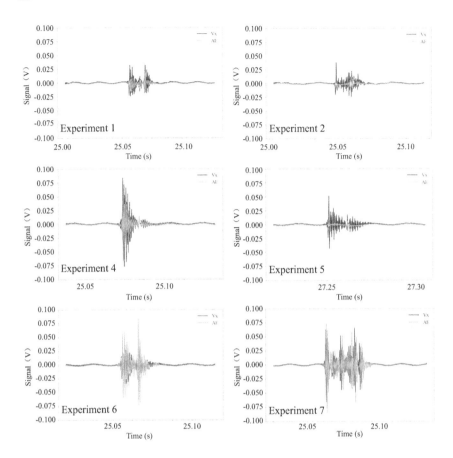

FIGURE 4.11   Vibration (Vx) signal and acoustic emission (AE) signal of a single cut.

materials, and processing parameters. Then, a clear understanding of the experimental steps and signal acquisition methods is covered. Finally, a preliminary analysis is conducted on the experimental results, and the relationship between the wear form of the coated surface milling cutter and the workpiece material is determined based on the wear image. The force signal, torque signal, vibration signal, and acoustic emission signal curves of a single cutting process under different working conditions are compared. The wear images obtained from this experiment, as well as signal data such as cutting force, torque, vibration, and acoustic emission, will be used for subsequent feature extraction and wear prediction.

4.1.3   Visual Wear Features and Signal Feature Extraction

4.1.3.1   Introduction

The amount of raw signal data in the face milling cutter wear experiment is very large. Direct input of raw signal data into the wear prediction model requires high computational complexity. Thus, it requires feature extraction. Therefore, this section mainly focuses on feature extraction, including wear features from images using visual methods and signal features from cutting force, torque, vibration, and acoustic emission signals based on time domain and frequency domain analysis. In addition, in order to eliminate noise interference, it is also necessary to preprocess the image and signal separately before feature extraction.

4.1.3.2 Visual wear feature extraction method

4.1.3.2.1 Image preprocessing

4.1.3.2.1.1 Image cropping The original photo (Figure 4.12) has a size of 1352 × 596, and the wear is mainly concentrated on the back face and secondary back face, so the first step is to crop the tool image. Using the upper edge of the secondary blade surface as the upper boundary of the cropped image, and the left vertex as the left boundary of the cropped image, the cropped image size is 320 × 80 (Figure 4.13).

4.1.3.2.1.2 Image grayscale Color images are obtained by overlaying the grayscale of three RGB channels, while for tool wear images, color information is not required. Therefore, the image can be grayed first to filter redundant color information and reduce the computational cost of subsequent feature extraction, as shown in Figure 4.14.

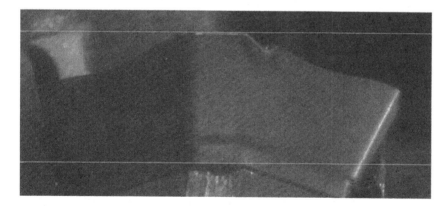

FIGURE 4.12 Original tool image.

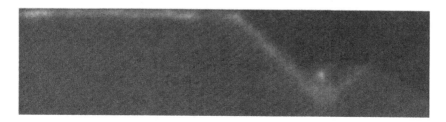

FIGURE 4.13 Cropped tool image.

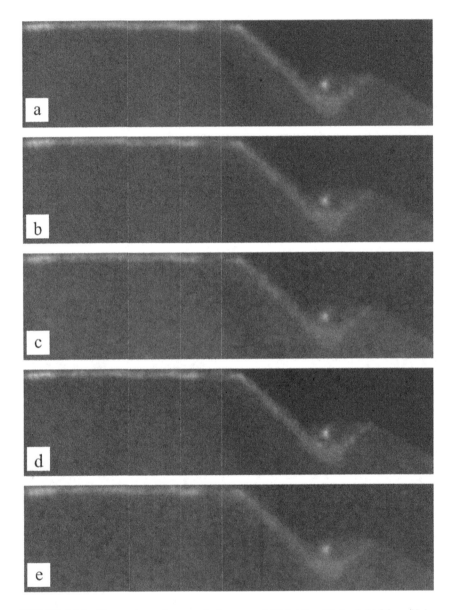

FIGURE 4.14 Comparison of grayscale images. a) Weighting Method; b) Mean Value Method; c) Single Channel R; d) Single Channel G; e) Single Channel B.

Common grayscale methods include three-channel mean method, three-channel weighting method, and single-channel method. The weighting method (Equation 4.1) refers to adding the grayscale values of different channels according to a certain coefficient ratio to obtain the final grayscale value. Due to the different sensitivities of the human eye to the three primary colors of red, green, and blue, the commonly used weighting coefficient is $w_R = 0.3$, $w_G = 0.59$, $w_B = 0.1$. The mean method (Equation 4.2) directly takes the average of the grayscale values of three channels as the final grayscale value.

$$GreyScale_{weighted} = w_R \times G_R + w_G \times G_G + w_B \times G_B \qquad (4.1)$$

$$GreyScale_{average} = \frac{G_R + G_G + G_B}{3} \qquad (4.2)$$

In the equation, G_R, G_G, G_B are the grayscale values of the three channels R, G, B. w_R, w_G, w_B are the weighting coefficients of the three channels, respectively.

Figure 4.14 shows the comparison of the effects of different grayscale methods. It can be seen that the boundaries of each region in the single-channel G grayscale image are clearer, so this method is selected for grayscale processing.

4.1.3.2.1.3 Image denoising and filtering methods Median filtering is a commonly used nonlinear filtering method that is suitable for processing salt and pepper noise and has the advantage of preserving edge clarity. The principle of median filtering is to use an odd-sized window to slide-scan the image. For the center position of the window, the filtered value is the median of all values within the window. Figure 4.15 shows a window size of 3×3, 5×5, and 7×7

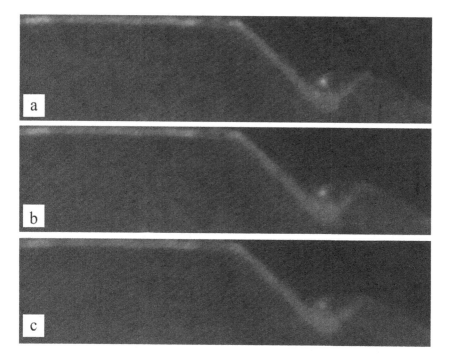

FIGURE 4.15   Images after median filtering. a) Window Size 3×3; b) Window Size 5×5; c) Window Size 7×7.

after median filtering. It can be seen that when the window is larger, its smoothing effect is better, but the clarity of the worn area is also slightly reduced. Therefore, using a 5 × 5 window is most suitable for median filtering.

Gaussian filtering is a commonly used linear filtering method, which uses a two-dimensional Gaussian distribution function to perform local convolution on the image. For filtering centered on position (i, j), the Gaussian function G_{ij} used is as follows (Equation 4.3), where σ_x and σ_y are the standard deviations of Distribution x and y, respectively:

$$G_{ij}(x, y) = e^{-\frac{(i-x)^2 + (j-y)^2}{2\sigma_x \sigma_y}} \tag{4.3}$$

Due to the fact that this function has a value of 1 when $x = i$, $y = j$ and tends to approach 0 when x, y is far away, the essence of its filtering is to use neighborhood gray value weighting to correct the gray value of the position (i, j), and the weight value is a Gaussian function value. Therefore, the weight of points closer to each other is higher. Figure 4.16 shows the results of Gaussian filtering. Compared to the median filtering results, the wear boundary clarity is insufficient and is not suitable for denoising this image. Therefore, the filtering method used here is median filtering.

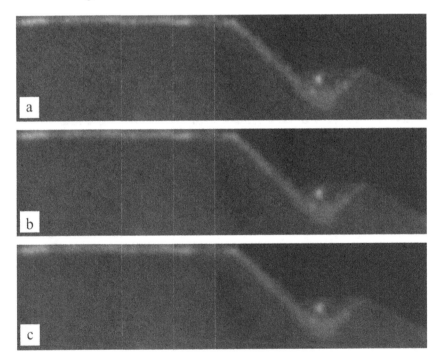

FIGURE 4.16 Images after Gaussian filtering. a) Convolution Kernel Size 3 × 3; b) Convolution Kernel Size 5 × 5; c) Convolution Kernel Size 7 × 7.

4.1.3.2.2 Wear edge extraction and region recognition

4.1.3.2.2.1 Binary wear region identification method Binarization can convert grayscale images into black-and-white images based on a set threshold for region recognition. For the wear image of the face milling cutter, it can be divided into three areas based on the grayscale value: background area, wear area, and tool area. The background area has a lower gray level, while the wear area has a higher gray level. In order to reduce the overall brightness error of different tool images, the binarization threshold can be selected through adaptive methods. That is, the maximum gray value in a certain area at the bottom of the secondary tool surface is taken as the binarization threshold.

We extract the binary results for the two situations of less tool wear and greater tool wear for analysis. Figure 4.17 shows the binarization results of the tool images after the second and 31st cuts. It can be seen that this method has a certain ability to recognize wear areas. However, due to the presence of high gray value areas on the rear tool surface, it is easy to be mistakenly recognized as wear areas. Some wear edge gray values do not increase significantly, which cannot be recognized by this method.

4.1.3.2.2.2 Edge extraction method based on Sobel operator The Sobel operator is 3×3 convolution factors (Figure 4.18), which can be convolved in both horizontal and vertical directions to obtain the grayscale gradient values in that direction. The values in both directions are combined to determine the size and direction of the grayscale gradient vector at this location, thus obtaining a grayscale gradient map. Due to the significant changes in grayscale at the boundary and the high absolute value of the gradient, the location of the boundary can be found by the grayscale gradient map obtained through Sobel convolution.

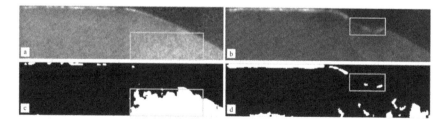

FIGURE 4.17 Images of tool after binarization. a) The grayscale tool image of second cutting; b) The grayscale tool image of 31st cutting; c) The binary tool image of second cutting; d) The binary tool image of 31st cutting.

(a)

-1	0	+1
-2	0	+2
-1	0	+1

(b)

+1	+2	+1
0	0	0
-1	-2	-1

FIGURE 4.18 Sobel convolution operator. a) Horizontal Sobel convolution operator; b) Vertical Sobel convolution operator.

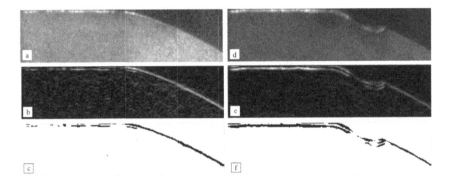

FIGURE 4.19 Edge recognition results of Sobel convolution. a) The grayscale image of second cutting; b) The grayscale gradient image of second cutting; c) The gradient binarization image of second cutting tool; d) The grayscale image of 31st cutting; e) The grayscale gradient image of 31st cutting; f) The gradient binarization image of 31st cutting.

Similarly, the Sobel edge detection results are extracted for analysis in two cases of less tool wear and greater tool wear. Figure 4.19 shows the grayscale gradient maps obtained by Sobel operator convolution of the tool images after the second and 31st cutting, as well as the results of binarization of the gradient maps. It can be seen that compared to direct binarization, the Sobel operator convolution has better edge recognition performance and is less affected by noise interference. However, the edges it identifies are thicker. In order to obtain more accurate wear, this recognition result still needs to be optimized.

4.1.3.2.2.3 Edge extraction method based on Canny algorithm   The Canny algorithm is also based on image grayscale gradient for edge detection. Compared to the Sobel operator, the Canny algorithm adds non-maximum suppression, which preserves only the maximum value in areas with large and slowly growing gradients. Therefore, it can refine the detected edges and more accurately locate the position of the edges. On this basis, the Canny algorithm has dual thresholds that can filter the detected gradient maximum edges again. For edges with gradients greater than a high threshold, they are identified as strong edges, and all are retained. For edges with gradients between two thresholds, they are identified as weak edges and only retained at weak edges where strong edges intersect. For edges with gradient values less than the low threshold, they are identified as errors caused by noise, and all are filtered. Therefore, the Canny algorithm can more effectively filter noise and obtain narrower and more accurate edges.

Similarly, the Canny edge detection results are extracted for analysis in two situations: less tool wear and greater tool wear. Figure 4.20 shows the grayscale image of the tool and the results of edge detection using the Canny algorithm.

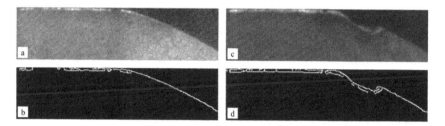

FIGURE 4.20 Edge recognition results of Canny method. a) The gray image of the second cutting; b) The edge extraction of the second cutting; c) The gray image of the 31st cutting; d) The edge extraction of the 31st cutting tool.

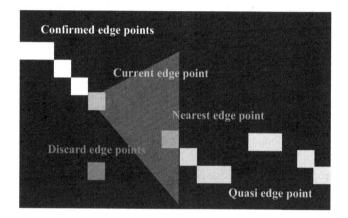

FIGURE 4.21 Schematic diagram of the edge repair algorithm.

Overall, the Canny algorithm can achieve good single-pixel accuracy in edge detection, especially for upper-edge detection. The detection accuracy of the outer contour edge position of the tool is relatively high. However, there are still some breakpoints that need to be further improved in terms of continuity. In addition, there are also a few non-real edge errors between the upper and lower edges, especially when the upper and lower wear edges of the auxiliary rear cutting surface are closer to the expected uneven distribution of brightness. Due to the importance of detecting the accuracy and continuity of the upper and lower edges for wear recognition, it is necessary to extract and repair the upper and lower edges.

Figure 4.21 is the schematic diagram of the edge repair algorithm.

Upper and lower edge extraction: First, scan the edges extracted by the Canny algorithm (Figure 4.22b) column by column from top to bottom. For each column of pixels, find the edge pixel point with the highest position. For the secondary flank, take the first row of pixels in the image to obtain the pseudo upper edge (Figure 4.22c1). Similarly, extract the bottom most edge pixel as the pseudo bottom edge (Figure 4.22c2).

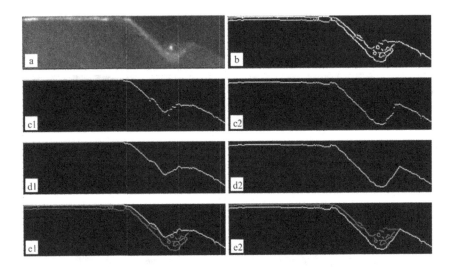

FIGURE 4.22 Results of edge extraction and repair. a) The gray image; b) Canny algorithm edge recognition result; c1) Pseudo upper edge; c2) Pseudo lower edge; d1) Corrected upper edge; d2) Corrected lower edge; e1) Corrected upper edge and Canny algorithm edge recognition result comparison; e2) Comparison of the corrected bottom edge and Canny algorithm edge recognition results.

Edge repair: Scan the pseudo edge from left to right, and the pseudo edge pixels in the first column are directly defaulted as confirmed edge points. During the scanning process, for each current pixel point, a triangle search is performed on the right pseudo edge point at a certain slope to find the nearest pseudo edge point, connect it for edge repair, and then use the nearest edge point as the current edge point to repeat the search-and-repair process on the right pseudo edge point. In addition, edge points located beyond the search area are discarded. Considering the shape of the tool and the distribution of edge points, for upper edge repair, the upper slope of the triangular search area is 1, and the lower slope is 1.5. For the repair of the lower edge, the upper slope of the triangular search area is 1, and the lower slope is 2. The repaired upper and lower edges are shown in Figure 4.22d1 and d2.

Figure 4.22e1 and e2 shows the results of upper and lower edge extraction and correction, respectively, compared with the initial edges obtained by the Canny algorithm. It can be seen that the repaired edges are more continuous and complete and can eliminate the interference of some discrete points, which is conducive to the identification of subsequent wear areas.

Wear area recognition: Use the upper edge extracted from the image of the unworn tool as the initial edge. Therefore, for the worn tool image, the area between the initial edge and the upper edge is a concave area, where the tool wear forms are concave and missing. The area between the upper and lower edges is the coating detachment zone, where the wear form is the coating detachment on the tool surface, but there is still the main material of the tool

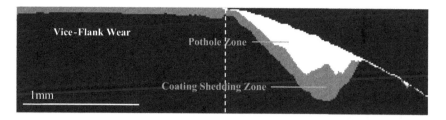

FIGURE 4.23 Result of wear recognition.

inside that has not yet fallen off. In comparison, pits are a more severe form of wear and have a greater impact on machining quality. The result can be seen in Figure 4.23.

4.1.3.2.3 Wear feature extraction

Pits and coating detachment are common forms of tool wear. End milling cutter wear is typically characterized by pits and missing edges, and turning tool wear is typically characterized by surface coating detachment. Therefore, the extraction of wear features should comprehensively consider these two forms of wear.

Based on the identified wear area map, the maximum wear width, average wear width, and wear area of the auxiliary and rear cutting surfaces can be extracted separately. For two different forms of wear, the wear amount is added using the weighted method. Due to the complete wear of the pits, their weighting coefficient is set to 1. Although coating detachment causes mild wear compared to pits, it also has a significant impact on the lifespan and processing quality of the coated surface milling cutter. Moreover, coating detachment can cause the main material of the tool to lose protection and friction with the workpiece during processing, accelerating its wear. Therefore, the weighting coefficient of the coating detachment zone is set to 0.8. However, the value of this coefficient does not affect the application of subsequent wear prediction methods. It can also be considered from other perspectives, by calculating the two wear forms separately or by adding equal weights to obtain the final wear amount.

Figure 4.24 shows the wear features extracted from the tool wear image in the previous experiment. The overall wear of the auxiliary rear blade surface is relatively small, with a maximum wear width of about 0.08 mm. As the wear frequency increases, the wear amount in this area increases rapidly in the early stage and then stabilizes. The wear of the rear blade surface is relatively large, with a maximum wear width greater than 0.4 mm. As the number of cuts increases, the wear amount in this area increases slightly in the early stage, but gradually accelerates in the later stage. Considering that the direct contact between the cutting surface and the material to be cut has a significant impact on the machining quality, more attention is also paid to the wear of the cutting surface in subsequent wear prediction. It is worth mentioning that the wear curve of face milling cutters is significantly different from that of end milling cutters,

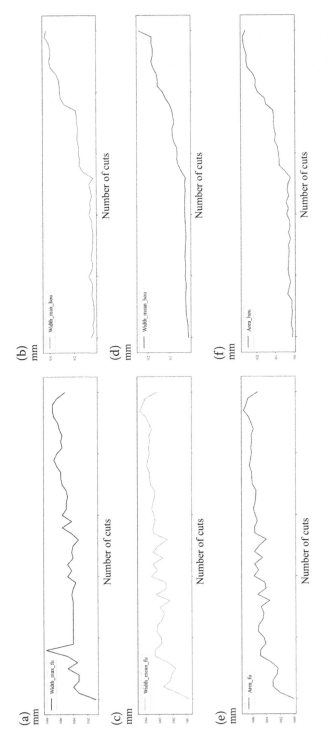

FIGURE 4.24 Curve of wear features—cutting times. a) Maximum wear width of the secondary flank surface; b) Average wear width of the secondary flank surface; c) Wear area of the secondary flank surface; d) Maximum wear width of the flank surface; e) Average wear width of the flank surface; f) Wear area of the flank surface.

and there is no obvious distinction between the three wear states. Therefore, the wear problem of face milling cutters is more suitable as a regression prediction problem rather than a classification problem.

4.1.3.3   Signal feature extraction

4.1.3.3.1   Signal preprocessing

4.1.3.3.1.1 Signal filtering   Due to the complex cutting environment, the raw signal data collected from sensors usually contains interference from non-cutting factors and environmental noise. Therefore, it is necessary to filter it, retain the wavelet related to wear, and eliminate the stable or high-frequency noise wavelet of interference.

To determine the filtering method and filtering threshold, Fourier transform can be performed first, and the signal spectrum can be observed as the number of cuts increases. Extract five equal portions of data from all wear signals in the experiment for fast Fourier decomposition, and compare the changes in amplitude values of each frequency sub wave during the wear growth process. According to the distribution of wavelet after Fourier decomposition, signals can still be divided into two types: the first type is force and torque signals; the second type is vibration and acoustic emission signals.

For the first type of signal, taking the cutting force signal in the x-direction as an example, divide the cutting signal into five equal parts and obtain its spectral distribution map after fast Fourier decomposition (Figure 4.25). It can be seen that the main wavelet of the signal is distributed in the frequency band less than

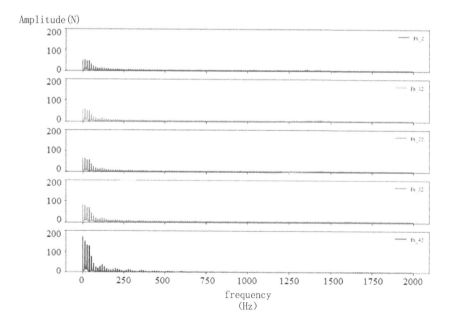

FIGURE 4.25   Spectrum of Fx signal after FFT.

250 Hz. As the wear amount increases, the amplitude value of the low-frequency sub wave gradually increases, and the amplitude increase is also significant when the wear is large in the later stage. Therefore, low-pass filtering can be used to eliminate high-frequency noise interference and preserve the core band related to wear, and the filtering threshold can be set to 250 Hz.

In addition, the force signal of a single cutting process can be divided into two parts: the cutting stage and the tool retraction stage. During the cutting stage, the blade comes into contact with the workpiece, causing the workpiece material to be cut off and resulting in a high cutting force. During the retraction stage, the blade leaves the workpiece, but there is still some residual force on the workpiece. The signals in both stages are subject to high-frequency noise interference before filtering, and interference needs to be eliminated. However, the force signal in the cutting stage can better reflect the state of the tool processing process, so its main waveform needs to be preserved.

Figure 4.26 shows the comparison of the three directional force signals and torque signals generated by a single cutting process before and after filtering.

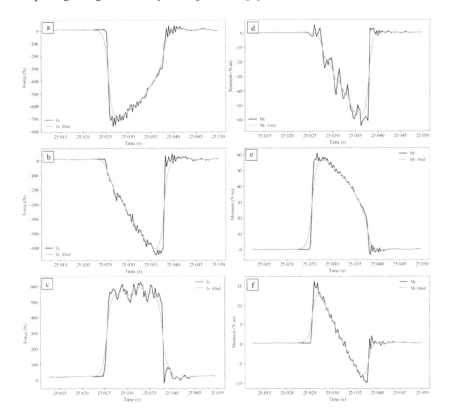

FIGURE 4.26 Signal filtering result of one cut. a) *x*-direction force signal; b) *y*-direction force signal; c) *z*-direction force signal; d) *x*-direction torque; e) *y*-direction torque; f) *z*-direction torque.

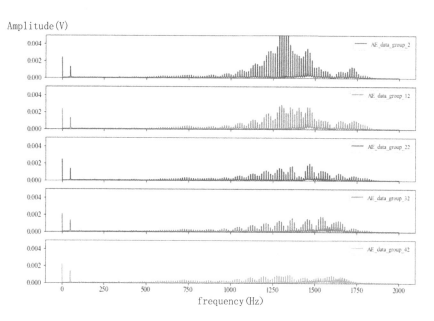

FIGURE 4.27 Spectrum of AE signal after FFT.

After filtering, the signal waveform during the cutting stage is highly preserved, while high-frequency noise and signal fluctuations during the tool retraction stage are filtered.

The same method can be used to filter and determine the threshold for vibration and acoustic emission signals. Taking acoustic emission signals as an example, Figure 4.27 shows the spectral distribution of acoustic emission signals in the same five sets of cutting processes after fast Fourier decomposition. Compared to force signals, the main wavelet distribution of acoustic emission signals is in the higher frequency range. As tool wear increases, the amplitude value of this frequency range gradually decreases. In addition, there are stable low-frequency waves caused by the environment, which do not change with the tool wear state and belong to unrelated wavelets that need to be filtered. Therefore, for such signals, a bandpass filter can be selected, with thresholds set to 700 Hz and 1800 Hz, while filtering low-frequency environmental waves and high-frequency noise waves.

Figure 4.28 shows the vibration signals and acoustic emission signals before and after filtering. According to the curve information in the figure, low-frequency environmental waves in the non-cutting stage were successfully filtered. However, due to the sampling frequency of 4000 Hz, the maximum wavelet frequency after Fourier decomposition is 2000 Hz, and the core wavelet frequency in the cutting stage is also high. There is a section where high-frequency noise intersects with the main wavelet frequency range, and its high-frequency noise filtering effect is limited.

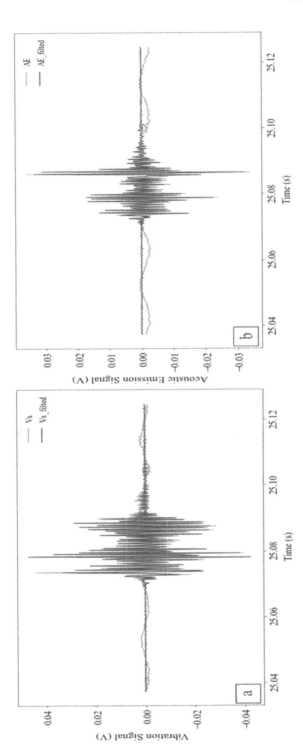

FIGURE 4.28 Signal filtering result of one cut. a) Vibration signal; b) AE signal.

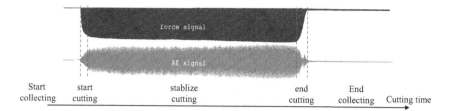

FIGURE 4.29 Force signal and AE signal collected during cutting.

4.1.3.3.1.2 Signal clipping   The collection of sensor signals starts before the start of cutting and stops after the end of cutting. Taking the x-direction force signal and acoustic emission signal as an example, Figure 4.29 shows the complete force and acoustic emission signals collected from the first cutting after filtering. It can be seen that during the initial cutting period, as the trajectory of the blade in contact with the workpiece gradually increases, the two signals grow rapidly. In the stable cutting stage, the length of the single cutting path is consistent, and the amplitude of the force signal and acoustic emission also changes slowly and stably. At the end of the cutting process, there is also a rapid decrease in signal amplitude. Therefore, it is necessary to crop the complete signal and extract the stable cutting stage signal to ensure the consistency of the signal during the cutting process, which is beneficial for subsequent analysis. At the same time, due to different signal acquisition software, there is also a time difference in the start collection time, and it is necessary to use the time point where the amplitude of each signal increases for time axis alignment.

When the signal is not cut, it fluctuates in a small range near the zero point. A threshold that is higher than the fluctuation range can be set. By scanning the complete signal from left to right, the time point at which the first amplitude of the signal exceeds the threshold can be found as the time point at which cutting begins. This time point is also used for time axis alignment between signals. At this time point, shift backwards for a period of time (i.e., skip the rapid growth stage of the signal at the beginning of cutting), and use it as the starting position for cutting. The signal time length after cutting $l_{cutted\ signal}$ is fixed, which is determined by the length of the workpiece feed direction $l_{workpiece}$, sampling frequency f, and feed rate v. Here, we round up to 4 cm as the corresponding length for the start and end stages of cutting. The fixed signal length after clipping is calculated as shown in Equation 4.4.

$$l_{cutted\ signal} = \frac{l_{workpiece} - 4\ cm}{v} * f \qquad (4.4)$$

Figure 4.30 shows the force, torque, vibration, and acoustic emission signals during the stable cutting stage after cropping and filtering. At this point, signal preprocessing has been completed. However, due to the long cutting time and high sampling frequency, the signal quantity after cutting is still huge, and direct

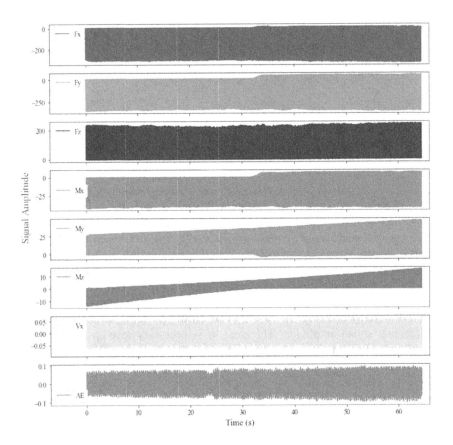

FIGURE 4.30 Sensor signal after cutting.

use for wear prediction requires significant computational performance. Therefore, it is necessary to extract features from the preprocessed signal.

4.1.3.3.2 Feature extraction

The feature extraction methods adopted in this chapter are divided into two types: time domain analysis and frequency domain analysis. Time domain analysis is the most intuitive feature extraction method, which essentially performs statistical analysis of signals in the time domain. Frequency domain analysis is a statistical analysis of the signal spectrum obtained based on Fourier transform from a frequency perspective.

Discrete Fourier transform (DFT) decomposes the original time domain signal into multiple sinusoidal wavelets of different frequencies and extracts the amplitudes of these wavelets as frequency domain information to obtain a spectrum of the signal. It is commonly used for analyzing periodic time series signals. The principle equation is shown in Equation 4.5, where an is the amplitude of the decomposed sinusoidal wavelet, N is a positive integer constant, and its value is related to the sampling frequency. For real number signals, the imaginary part after decomposition is 0, and only the real part can be considered. Fast Fourier transform (FFT), on the other hand, is an improved method that utilizes some symmetry and periodicity rules in its calculation process to reduce the amount of repetitive operations in complex number calculations and improve computational efficiency based on butterfly transform. Therefore, the results of FFT and DFT are the same, but the computational complexity is significantly reduced.

$$f(x) = \sum_{n=0}^{N-1} a_n e^{-j\frac{2\pi xn}{N}} = \sum_{n=0}^{N-1} a_n \left(\cos \frac{2\pi xn}{N} - j \sin \frac{2\pi xn}{N} \right) \tag{4.5}$$

The statistical analysis features used include maximum, minimum, mean (Equation 4.6), median, standard deviation (Equation 4.7), root mean square (Equation 4.8), skewness (Equation 4.9), kurtosis (Equation 4.10), waveform factor (Equation 4.11), peak factor (Equation 4.12), pulse factor (Equation 4.13), and margin factor (Equation 4.14)—12 in total.

$$\bar{x} = \frac{1}{N} \sum x_i \tag{4.6}$$

$$x_{std} = \sqrt{\frac{1}{N} \sum (x_i - \bar{x})^2} \tag{4.7}$$

$$x_{rms} = \sqrt{\frac{1}{N} \sum x_i^2} \tag{4.8}$$

$$x_{skew} = \frac{1}{N} \frac{\sum (x_i - \bar{x})^3}{x_{std}^3} \tag{4.9}$$

$$x_{kurt} = \frac{1}{N} \frac{\sum (x_i - \bar{x})^4}{x_{std}^4} \tag{4.10}$$

$$x_w = \frac{x_{rms}}{\frac{1}{N} \sum |x_i|} \tag{4.11}$$

$$x_{pe} = \frac{\max |x_i|}{x_{rms}} \tag{4.12}$$

$$x_{pu} = \frac{\max |x_i|}{\frac{1}{N} \sum |x_i|} \tag{4.13}$$

$$x_w = \frac{\max |x_i|}{\left(\frac{1}{N} \Sigma \sqrt{|x_i|}\right)^2} \qquad (4.14)$$

Therefore, these 8 signals extract 12 features in the frequency domain and time domain, respectively, resulting in a total of 192 extracted features. Drawing can obtain the curve of the variation of each feature value with the number of processing times. Represented by the force signal and acoustic emission signal in the x-direction, Figure 4.31 shows the time domain feature extraction results of these two signals in the first set of experiments. Figure 4.32 shows the frequency domain feature extraction results of these two signals in the first set of experiments, where the horizontal axis represents the number of processing times, and the vertical axis represents the signal feature values. In order to balance the impact of different features on the results and facilitate horizontal comparison, all feature values will be homogenized in the subsequent dataset preparation, reducing their distribution to 0–1. Therefore, the dimensions of the extracted feature values here can be ignored.

Due to the monotonic increase of tool wear with the increase of processing times, the trend of signal characteristics with the increase of processing times also reflects the impact of increased wear. From the trend of signal characteristics, the time domain characteristics of force signals have more obvious regularity, while the frequency domain characteristics of acoustic emission signals have more obvious regularity. This also proves the necessity of extracting features separately in the frequency domain and time domain. Overall, the extracted features can reflect the tool wear status to a certain extent, which is conducive to the establishment of subsequent wear prediction models. However, some features have information redundancy, which means that there is high similarity between some feature curves. In addition, there are also some features with low correlation, which means that the feature curves do not change significantly or have obvious regularity with the number of processing times. Therefore, it is necessary to screen the features.

4.1.3.4 Summary

This section extracts feature values from the experimental data, including visual wear feature extraction and signal feature extraction. For tool wear images, preprocessing such as image cropping, graying, filtering, and noise reduction are first performed. Then, after comparing some edge recognition and region recognition algorithms, it is confirmed to use the edge recognition method based on the Canny algorithm. The recognition results are improved to obtain the wear areas corresponding to the two wear forms of pits and coating detachment. Finally, the weighted method is used to obtain the maximum wear width of the secondary and rear tool surfaces, respectively. The average wear width and wear area are used as wear characteristics. For the force signal, torque signal, vibration signal, and acoustic emission signal obtained from the experiment, a total of 12 statistical features were extracted using time domain and frequency domain analysis

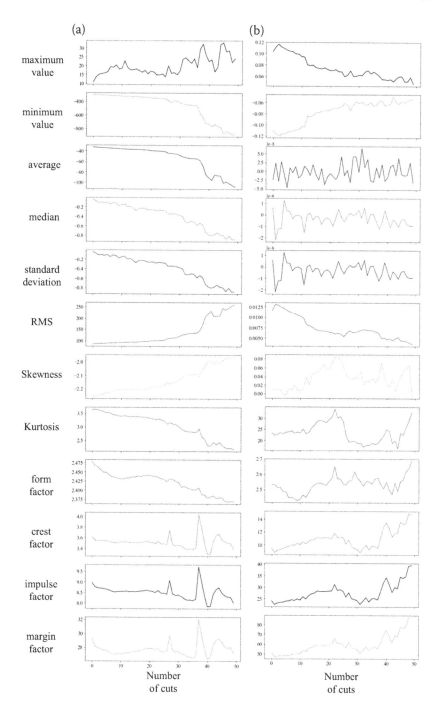

FIGURE 4.31 Time domain feature extraction results. a) Time domain feature extraction of Fx signal; b) Time domain feature extraction of AE signal.

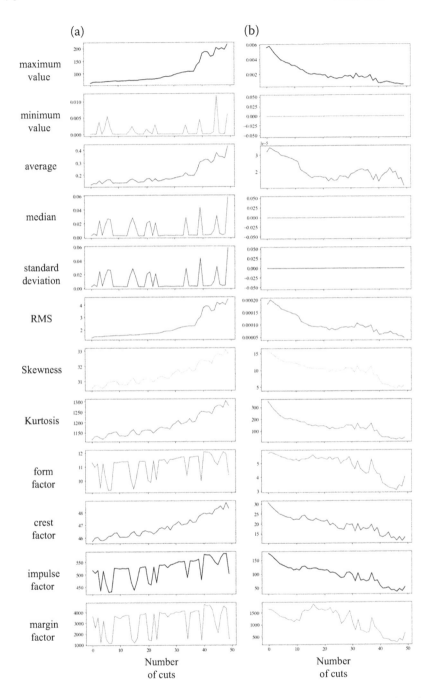

FIGURE 4.32 Frequency domain feature extraction results. a) Frequency domain feature extraction of Fx signal; b) Frequency domain feature extraction of AE signal.

methods, including maximum, minimum, mean, median, standard deviation, root mean square, skewness, kurtosis, waveform factor, peak factor, impulse factor, and margin factor. Finally, 192 signal features were obtained to lay the foundation for the establishment of subsequent wear prediction models.

4.1.4 Tool Wear Prediction Method Under Multiple Working Conditions

4.1.4.1 Introduction

The essence of the data-driven wear prediction method is to establish a regression model based on signal features. In order to improve the universality of the prediction model, this section uses a feature transfer method to solve the problem of insufficient samples in multiple working condition scenarios. On the basis of feature extraction in the previous section, this section first conducts dataset preparation and feature selection. Then the section compares and analyzes the wear prediction methods in a single working condition scenario. Finally, based on the feature transfer method, a wear prediction model suitable for multiple operating conditions is obtained and compared with the non-migrated model.

4.1.4.2 Dataset preparation

From the experimental data, it can be concluded that the wear of surface milling cutters is strongly related to the hardness of the workpiece. Different hardnesses of the workpieces result in different forms of wear, which is not suitable for comparison. Therefore, the wear prediction research in this chapter is based on scenarios where the same workpiece material is used but the processing parameters are different. When cutting stainless steel workpieces with higher hardness, the wear speed and form of the face milling cutter are faster and more severe. It can be seen that the wear of the face milling cutter when processing harder materials has more research value and depth. Therefore, the face milling cutter dataset used in this chapter selects the data collected from the first three groups of hard stainless steel workpieces processed under different operating conditions and parameters. Among them, the data obtained from the first group of experiments serves as the source dataset, while the data obtained from the second and third groups of experiments serve as the target dataset. The operating parameters are shown in Table 4.2.

TABLE 4.2
Processing Parameters of Face Milling Dataset

No	v_f (mm/min)	n (r/min)	a_p (mm)	f (mm/r)	Domain
1	80	800	0.9	0.1	Source
2	60	1200	1.2	0.05	Target 1
3	80	1000	1.2	0.08	Target 2

Perform feature extraction on the experimental data according to the method introduced in the previous section. Compared to the auxiliary rear cutting surface, the wear of the rear cutting surface is more severe and has a greater impact on machining quality. Moreover, the wear area has a more comprehensive representation of the wear situation compared to the maximum wear width. Therefore, the wear area of the rear cutting surface is selected as the core indicator to measure the wear amount, which is the target value of the wear prediction model. The 12 time domain features and 12 frequency domain features are extracted for each signal, resulting in a total of 192 signal features.

From Figures 4.31 and 4.32, it can be seen that there is still some noise fluctuation in the extracted features, especially in the wear feature curve, whose wear amount should monotonically increase with the increase of processing times. Therefore, it is necessary to post-process the feature curve. This chapter uses the Savitzky Golay filter to smooth each feature time curve.

The principle of this filter is to use a sliding window, apply the least squares method to polynomial fit the local curve, and determine the value of the center point of the window based on the fitted polynomial expression. The parameters that can be set are the size of the sliding window and the degree of the polynomial to be fitted. The window size should be greater than the degree of the polynomial to ensure that there is a unique optimal solution for the fitting. Considering the local monotonicity of the feature curve, its polynomial degree can be set to 1. To balance its filtering and smoothing effect with the preservation of the local trend of the curve, the window size is set to 7 data points. The specific fitting Equation 4.15 is as follows, where (X, Y) is the original data points within a window size of $2w - 1$. (X, Y_s) is the smoothed data points, J is the loss function, and θ is the smoothing curve parameter:

$$Y_s = X\theta \tag{4.15}$$

$$J = \frac{1}{2}(X\theta - Y)^T(X\theta - Y) \tag{4.16}$$

$$X = [x_{i-w}, \ldots \ldots, x_i, \ldots \ldots, x_{i+w}] \tag{4.17}$$

$$Y = [y_{i-w}, \ldots \ldots, y_i, \ldots \ldots, y_{i+w}] \tag{4.18}$$

Taking the wear characteristic curve as an example, Figure 4.33 shows the comparison of the results before and after filtering the wear amount in three sets of experiments. It can be seen that the smoothed wear curve has the same trend as the original curve, but local noise interference has been successfully filtered.

In addition, in order to enrich the sample data, this chapter uses the interpolation method to amplify the filtered data, expanding the feature data of

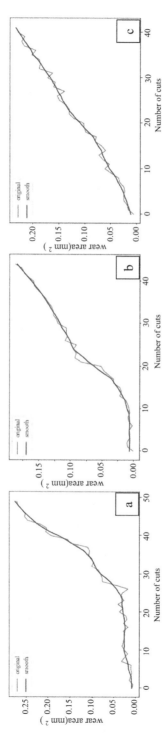

FIGURE 4.33 Results of Savitzky Golay filter. a) Experiment 1; b) Experiment 2; c) Experiment 3.

about 50 cuts in each group of experiments into about 300 sets of feature data, which is conducive to improving the accuracy of the model.

The total number of features obtained from signal feature extraction is 192. Due to the extraction of the same statistical features in both the time and frequency domains for each signal, there are inevitably some features that have weak correlation with wear. Therefore, preliminary screening is necessary to improve the accuracy of the wear prediction model and the computational efficiency of the feature transfer model. This section uses Pearson correlation analysis for preliminary feature selection and calculates the correlation coefficient between two features based on the variance and covariance of the feature distribution. The calculation method for the correlation coefficient r between feature Y and feature r is shown in Equation 4.19.

$$r(X, Y) = \frac{Cov(X, Y)}{\sqrt{Var(X)Var(Y)}} \qquad (4.19)$$

Based on the data from the experiment, the correlation coefficients between each feature and wear amount can be obtained. Considering the need to remove failed features after subsequent feature migration, the initial feature selection should leave redundancy. This chapter sets the threshold for feature filtering here to 0.6, which means removing features with a correlation coefficient less than 0.6 with wear, and the total number of features changes to 131.

Finally, three face milling cutter datasets were obtained, each containing approximately 300 sets of data, each containing 132 filtered signal features, and the wear area of the later tool face was used as the prediction object for wear amount. In the following text, in Section 4.1.4.3, these three datasets will be used for comparative analysis of wear prediction methods in single working condition scenarios. In Section 4.1.4.4, migration models from Dataset 1 to Datasets 2 and 3 will be established for wear prediction in multiple working condition scenarios.

4.1.4.3 Wear prediction method under single working condition scenarios

Machine learning methods and neural network methods have high accuracy in predicting the wear of end mills and turning tools under single working conditions, but there is relatively little research on the wear prediction of face mills. Therefore, this case study compares and analyzes the wear prediction effects of support vector machine regression, random forest (RF), and BP neural network on the same dataset. The prediction accuracy is evaluated using the determination coefficient (R Squared, R2) and mean square error (MSE). Evaluate functions such as root mean squared error (RMSE), and consider the model training time and prediction time to measure computational efficiency and resource loss.

4.1.4.3.1 Support vector machine algorithms

The most basic support vector machine (SVM) method is commonly used for linearly distinguishable binary classification problems. The principle is to find an

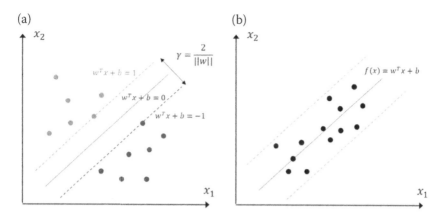

FIGURE 4.34 SVM and SVR. a) SVM; b) SVR.

optimal plane $w^T x + b = 0$ so that all points in the space can be divided into two sets, $D_1 D_2$ and D_2, and all points in set $D\,1$ satisfy $w^T x + b > 0$, and all points in set D_2 satisfy $w^T x + b < 0$. The optimization condition is to maximize the minimum distance from the points in the two sets to the hyperplane; hence, this optimal hyperplane is also known as the maximum interval hyperplane. Figure 4.34a shows a Schematic diagram of SVM classification principle, when the dimensions of the dataset are in two-dimensional case. Due to the synchronous scaling of parameters w and b, it can be considered that the closest point between the two sets and the hyperplane with the maximum distance is located on the plane $w^T x + b = \pm 1$. Therefore, the interval between them (margin) is defined as $\gamma = \frac{2}{\|w\|}$. Therefore, the optimization problem and constraints are as shown:

$$\max_{w,b} \frac{2}{\|w\|} = \min_{w,b} \frac{1}{2} \|w\|^2 \tag{4.20}$$

$$s.t.\ \ y_i(w^T x_i + b) \geq 1,\ i \in \mathbb{Z}^+,\ y_i \in \{1, -1\} \tag{4.21}$$

However, in reality, the distribution of data points often cannot satisfy strict separability, especially for data points distributed near boundaries. Therefore, a soft interval region is defined near the maximum interval hyperplane. For data points in this region, their classification errors can be ignored in optimization problems. Considering the computational complexity of optimizing non-continuous functions, hinge loss functions (Equation 4.22) are often used instead of 0–1 loss functions (Equation 4.23). To simplify expressions and facilitate calculations, slack variables ξ can be used, as defined in Equation 4.24.

$$0 - 1 \text{ loss: } l_{0-1}(x) = \begin{cases} 1, & \text{if } x < 0 \\ 0, & \text{if } x \geq 0 \end{cases} \tag{4.22}$$

$$\text{hinge loss: } l_h(x) = \max(0, 1 - x) \tag{4.23}$$

$$\xi_i = \max(0, 1 - y_i(w^T x_i + b)) \tag{4.24}$$

Therefore, the optimization problem can be modified as shown in Equation 4.25 and Equation 4.26, where C is the regularization constant and is one of the hyperparameters in the model that can be adjusted according to the target.

$$\min_{w,b} \frac{1}{2} \|w\|^2 + C \Sigma_i \xi_i \tag{4.25}$$

$$s.t. \ y_i(w^T x_i) \geq 1 - \xi_i, \ \xi_i \geq 0, \ i \in \mathbb{Z}^+ \tag{4.26}$$

Support vector regression (SVR) is a branch of SVM in the field of regression problems. The principle is: for the dataset $D = \{(x_i, y_i), i = 1,2,3 \ldots \ldots\}$, find a regression function $f(x) = w^T x + b$ to minimize the sum of the distances between the true values y_i and the regression values $f(x_i)$ of all data points. Therefore, when the function fitting results are good, for the new data x_{new}, it can be considered that its regression value can replace the true value, namely $f(x_{new}) \approx y_{new}$.

Similarly, regression problems can also define an interval with a width of 2ϵ near the hyperplane. For data points falling within this area, the error can be considered zero. For data points outside of the region i, the difference between $f(x_i)$ and y_i can be calculated. To maintain the continuity of the function, ϵ needs to be subtracted as the loss value. Similarly, relaxation variables ξ and $\hat{\xi}$ can be introduced, as defined in Equations 4.27 and 4.28.

$$\xi_i = \max(0, f(x_i) - y_i - \epsilon) \tag{4.27}$$

$$\hat{\xi}_i = \max(0, y_i - f(x_i) - \epsilon) \tag{4.28}$$

The optimization function of this regression problem can be transformed as shown in Equation 4.29 and 4.30.

$$\min_{w,b} \frac{1}{2} \|w\|^2 + C \sum_i \left(\xi_i + \hat{\xi}_i \right) \tag{4.29}$$

$$s.t. \ f(x_i) - y_i - \epsilon \le \xi_i, \ \xi_i \ge 0,$$
$$y_i - f(x_i) - \epsilon \le \widehat{\xi_i}, \ \widehat{\xi_i} \ge 0, \ i \in \mathbb{Z}^+ \tag{4.30}$$

Optimization problems with constraints can be transformed into unconstrained optimization problems using the Lagrange method. The function expression is shown in Equation 4.31, where α, α̂, β, β̂ are a non-negative Lagrange multiplier.

$$L(w, \ b, \ \alpha, \ \widehat{\alpha}, \ \beta, \ \widehat{\beta}) = \frac{1}{2}\|w\|^2 + C\sum_i \left(\xi_i + \widehat{\xi_i}\right) + \sum_i \alpha_i \left(f(x_i) - y_i - \epsilon - \xi_i\right)$$
$$+ \sum_i \widehat{\alpha_i}\left(f(x_i) - y_i - \epsilon - \widehat{\xi_i}\right) - \sum_i \beta_i \xi_i - \sum_i \widehat{\beta_i}\widehat{\xi_i}$$

$$\tag{4.31}$$

Apply the Karush Kuhn Tucker condition (KKT), we can obtain the expression of w (Equation 4.32). We calculate the mean of each data point to obtain the expression of b (Equation 4.33).

$$w = \sum_i (\widehat{\alpha_i} - \alpha_i)x_i^T \ \text{with} \begin{cases} if \ \xi_i > 0, \ \widehat{\xi_i} = 0 \text{: } \alpha_i = C, \ \widehat{\alpha_i} = 0 \\ if \ \xi_i = 0, \ \widehat{\xi_i} > 0 \text{: } \alpha_i = 0, \ \widehat{\alpha_i} = C \end{cases} \tag{4.32}$$

$$b = avg\left(y_i + \epsilon - \left(\sum_j \left(\widehat{\alpha_j} - \alpha_j\right)x_j^T\right)x_i\right) \tag{4.33}$$

For data with nonlinear distribution, the data points can be first projected onto a higher dimensional space to obtain $\varphi(x)$, and then classified using SVM or regressed using SVR. The optimization calculation method is the same, and using the kernel function $K(x, z) = \varphi(x) \cdot \varphi(z)$ can simplify the calculation of its inner product. The kernel function used in this chapter is the radial basis kernel function, and its Equation 4.34 is as follows, where σ and γ are constants:

$$K(x, z) = \exp\left(-\frac{\|x - z\|^2}{2\sigma^2}\right) = \exp(-\gamma\|x - z\|^2) \tag{4.34}$$

4.1.4.3.2 Random forest algorithm

A classification decision tree is an algorithm established based on the principle of binary trees to address classification problems. Its branch nodes are eigenvalue conditions, and the leaf nodes are the categories to which they belong. The generation of decision trees focuses on the selection of feature conditions, which can be achieved using ID3 algorithm, C4.5 algorithm, and CART algorithm. The feature splitting criteria used in these three algorithms are information gain, information gain rate, and Gini index, and their calculation methods and related definitions are as follows:

- **Information entropy H (H):** Used to measure the uncertainty of variable distribution. Assuming that the values of feature X are $x_1, x_2, \ldots \ldots, x_n$, and the probability of each value x_i is p, the calculation method for the information entropy of feature X is shown in Equation 4.35.

$$H(X) = -\Sigma_i p_i \log p_i \tag{4.35}$$

- **Conditional entropy $H(Y|X)$:** Used to measure the uncertainty of the value of variable Y when the value of variable X is known in Equation 4.36.

$$H(Y|X) = \Sigma_i p_i H(Y|X = x_i) \tag{4.36}$$

- **Information gain $g(Y, X)$:** Used to measure the degree of reduction in uncertainty of variable Y when X is known in Equation 4.37.

$$g(Y, X) = H(Y) - H(Y|X) \tag{4.37}$$

- **Information gain rate $g_r(Y, X)$:** Used to normalize the information gain based on it. It's shown in Equation 4.38.

$$g_r(Y, X) = \frac{H(Y) - H(Y|X)}{H(X)} \tag{4.38}$$

- **Gini index $G(X)$:** Similar to information entropy, used to measure the uncertainty of variable X. It's shown in Equation 4.39.

$$G(X) = \Sigma_i p_i(1 - p_i) \tag{4.39}$$

At present, the CART algorithm is more widely used compared to the other two algorithms. This algorithm adds pruning to the decision tree, which searches for the optimal subtree based on the loss function for the already generated decision tree, effectively improving the model's overfitting resistance.

Regression tree is the application of decision tree in continuous problems. For continuous variables, variance can be used to measure the degree of dispersion of their distribution, replacing the Gini index in the CART algorithm. To branch a regression tree, it is necessary to calculate the sum of variances for each subset

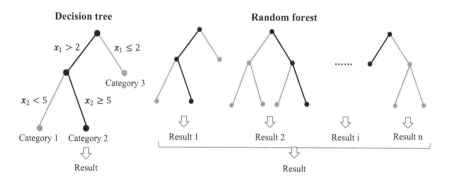

FIGURE 4.35 Decision tree and random forest.

under each segmentation situation, find the optimal segmentation feature and segmentation point, and repeat the branching steps until the stop condition is reached. The commonly used stopping iteration conditions include a minimum sample size limit for leaf nodes, a maximum depth limit for regression trees, and a minimum error threshold. The control of stop conditions can also achieve the effect of pruning, avoiding the generation of too many leaf nodes, and the problem of overfitting. Finally, the predicted value of the regression tree is the mean label value of all samples contained in the leaf node.

RF is an ensemble learning method proposed on the basis of decision trees, which essentially involves learning and predicting multiple unrelated decision trees separately and then summarizing them to obtain the final results. For classification problems, the final result takes the mode of the results obtained from all classification trees. For regression problems, the final result is taken as the mean of the results obtained from all regression trees.

This algorithm has two random aspects. One is to randomly select a portion of samples when establishing each decision tree. The second is to randomly select some features when adding branch nodes and then optimize them to determine the splitting conditions. These two random aspects make this algorithm have better resistance to noise and can handle a large number of features, thus reducing the requirements for feature selection.

Figure 4.35 shows decision tree and random forest.

4.1.4.3.3 Backpropagation neural network algorithm

Backpropagation neural network (BPNN), as one of the most fundamental and widely used neural network algorithms, is modeled after biological neural networks and consists of multiple layers of connected neurons (Figure 4.36): The first layer is the input layer, and the number of neurons depends on the number of input features. The last layer is the output layer, and its number of neurons depends on the final desired prediction result form. The middle layer, also known as the hidden layer, is between the input layer and the output layer. The number of layers and neurons in the middle layer can be adjusted according to the application scenario.

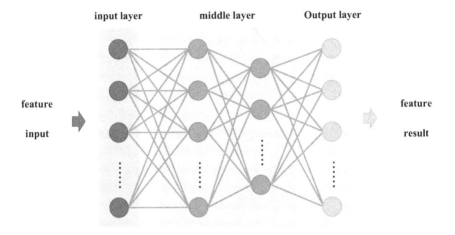

FIGURE 4.36 Backpropagation neuron network.

For a certain neuron in the *i-th* layer, its input value x_i is equal to the output of each neuron in the previous layer y_{i-1}. The output value is obtained by multiplying x_i by the weight matrix w_i and then using the activation function *f*:

$$x_i = y_{i-1} \tag{4.40}$$

$$y_i = f(w_i^T x_i) \tag{4.41}$$

The activation function of biological neurons is a 0–1 step function. For the convenience of calculation and differentiation, *sigmoid* or *relu* functions are often used as substitutes in neural network algorithms. The three activation function Equations 4.42–4.44 and curves are as follows:

$$f_{0-1}(x) = \begin{cases} 1, & if \ x < 0 \\ 0, & if \ x \geq 0 \end{cases} \tag{4.42}$$

$$f_{sigmoid}(x) = \frac{1}{1 + e^{-x}} \tag{4.43}$$

$$f_{relu} = \max(0, x) \tag{4.44}$$

After obtaining the prediction result y_{pre}, the calculation error can be compared with the standard value y_{real}. This chapter uses the mean square deviation E_{mse}, and its calculation Equation 4.45 is as follows:

$$E_{mse} = \frac{1}{2} \left\| y_{pre} - y_{real} \right\|^2 = \frac{1}{2} \sum_i (y_{pre,i} - y_{real,i})^2 \qquad (4.45)$$

The initial weight matrix w is generated by random numbers. To reduce errors, it is necessary to update and iterate the weight matrix. This update is based on the gradient descent method, the update amount of w (Δw) is obtained by taking the partial derivative of the error E_{mse} to w, and the calculation Equation 4.46 is as follows, where η is the learning rate:

$$\Delta w = -\eta \frac{\partial E_{mse}}{\partial w} = -\eta \frac{\partial E_{mse}}{\partial y} \frac{\partial y}{\partial w} \qquad (4.46)$$

The number of iterations or error accuracy can be set as stopping conditions to ultimately obtain a more accurate neural network model for prediction.

4.1.4.3.4 Analysis of prediction results

After shuffling the order of each group of experimental data, take 70% as the training set and the other 30% as the test set. Select the parameters of the model based on the training set data from Experiment 1. The parameter selection of SVM model and RF model adopts the grid search cross-validation method. First, the parameter range is defined, and then each group of parameters is validated and calculated on the training dataset. That is, the training dataset is divided into multiple parts, one of which is taken as the test set, and the remaining is used as the training set. Finally, the comprehensive results are scored to find the best parameter combination. The BPNN model adopts manual parameter adjustment to ultimately obtain the optimal parameters of these three models in the scenario of predicting surface milling cutter wear Tables 4.3–4.5.

This chapter uses evaluation functions such as the coefficient of determination (R^2), mean square error (MSE), and RMSE to compare the prediction results of

TABLE 4.3

Parameters of SVR Model

kernel	C	γ	ϵ
rbf	100	0.001	0.001

TABLE 4.4

Parameters of RF Model

criterion	n_estimators	max_depth	min_samples_split
mae	150	100	2

TABLE 4.5
Parameters of BPNN Model

input size	hidden size_1	hidden size_2	output size	num epochs	learning rate
131	64	16	1	100000	0.1

various models under single operating conditions. The evaluation function calculation Equations 4.47–4.48 are as follows, where y_p is the predicted value, y_r is the true value, and $\overline{y_r}$ is the mean of the true value:

$$R^2 = 1 - \frac{\sum_{j=1}^{n} (y_{p,j} - y_{r,j})^2}{\sum_{j=1}^{n} \left(y_{p,j} - \overline{y_r} \right)^2} \in [0, 1] \tag{4.47}$$

$$MSE = RMSE^2 = \frac{1}{n} \sum_{j=1}^{n} (y_{p,j} - y_{r,j})^2 \tag{4.48}$$

According to the equation, when R^2 approaches 1, RMSE and MES approach 0, and the model's prediction results become more accurate. Table 4.6 shows the prediction results of three methods on three datasets. For the accuracy of the results, all three prediction methods achieved good results in a single operating scenario, with the determination coefficient R^2 greater than 0.999 and RMSE less than 0.1, with RF being the best. For the consumption of computing resources, BPNN has the longest training time and SVM has the shortest training time, but the prediction time of the three models is similar. Overall, both SVM and RF meet the performance requirements in a single operating scenario.

TABLE 4.6
Prediction Results of Single-Condition Scenarios

Forecasting Methods	Dataset	R^2	RMSE	MSE	Training Time	Forecasting Time
SVM	1	0.99990	0.03262	0.00106	15 ms	324 ms
	2	0.99979	0.04644	0.00216	8 ms	343 ms
	3	0.99987	0.02180	0.00048	20 ms	300 ms
RF	1	0.99994	0.01018	0.00010	3.9 s	337 ms
	2	0.99997	0.01462	0.00021	3.4 s	343 ms
	3	0.99994	0.00916	0.00008	3.2 s	312 ms
BPNN	1	0.99981	0.03747	0.00140	65.7 s	338 ms
	2	0.99942	0.04415	0.00195	71.0 s	405 ms
	3	0.99974	0.01921	0.00037	69.5 s	334 ms

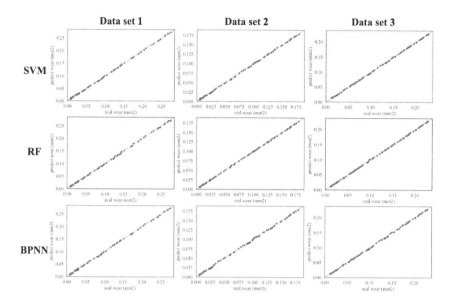

FIGURE 4.37 Prediction results of single-condition scenarios.

Figure 4.37 shows the predicted results curves of three methods on three datasets, where the horizontal axis represents the actual wear value, the vertical axis represents the predicted wear value, and the red line represents the function $y = x$, which facilitates a more intuitive observation of the predicted results. It can be seen that almost all the predicted points fall on the straight line $y = x$, indicating that the predicted results are very close to the true values, proving the extremely high accuracy of these three prediction methods in single working condition scenarios.

4.1.4.4 Method of feature transfer in multiple working condition scenarios

The previous section has verified the good predictive ability of SVM algorithm, RF algorithm, and BPNN algorithm in single operating conditions. However, models with good predictive performance in one operating condition usually cannot be applied in other operating conditions. For example, using the data from Experiment 1 as the training set, and the data from Experiment 2 and Experiment 3 as the test set, the predicted results are as follows. Considering the long calculation time of the BPNN algorithm, only the SVM algorithm and the RF algorithm are compared in Table 4.7.

From the data in the above table, it can be seen that the RF algorithm has better generalization ability, but overall, the prediction accuracy of both methods is significantly reduced compared to a single operating scenario. Figure 4.38 shows the comparison curves between the wear prediction results of different test

TABLE 4.7
Prediction Results of Multi-Condition Scenarios

Forecasting Methods	Training Dataset	Test Dataset	R^2	RMSE	MSE
SVM	1	2	-3.54066	9.32387	86.93463
	1	3	-0.52257	1.28939	1.66252
RF	1	2	0.46832	4.72062	22.28422
	1	3	0.29289	0.66918	0.44780

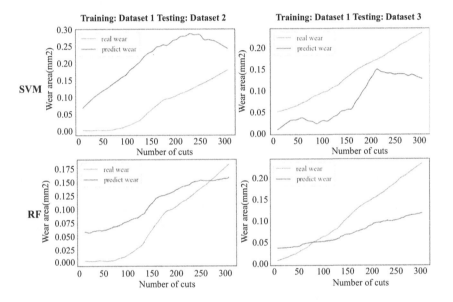

FIGURE 4.38   Prediction results of test set for multi-condition scenarios.

sets and the actual wear values under different prediction models. Although the predicted value curve has the same trend as the actual value curve, there is still a significant gap. This also demonstrates the coexistence of data similarity and specificity in scenarios with multiple operating conditions, proving the feasibility of transfer learning.

4.1.4.4.1   Feature transfer method

The features extracted based on historical tool data are the source features, while the features extracted based on new tool data are the target features. In this chapter, the source features are extracted from the Experiment 1 dataset after preliminary screening, while the target features are extracted from the Experiment 2 and Experiment 3 datasets after the same screening.

The core of feature transfer is to use the early data of the target feature dataset to construct a transfer relationship with the early data of the source dataset, modify the source feature dataset to better match the distribution of the target features, and then use the migrated features to train the prediction model, aiming to improve the prediction accuracy of the prediction model on the target dataset.

First, a feature transfer model is established using pre wear data, which is the first 150 sets of data. The maximum and minimum normalization transformations are performed on the source domain features and target domain features, respectively, so that the distribution of each feature is between 0 and 1. The normalization method for feature Δ is shown in Equation 4.49.

$$f(X_i) = \frac{X_i - \min X_i}{\max X_i - \min X_i} \tag{4.49}$$

After transformation, a one-to-one linear transfer model is constructed for each corresponding feature in the source and target domains. The linear transfer model equation for feature X_i is shown in Equation 4.50, where f_t, f_s are feature normalization processing functions for the target domain and source domain, respectively. a_i, b_i are the migration parameters for this feature.

$$X_{tr,i} = f_{t,i}^{-1}(a_i^T f_{s,i}(X_{s,i}) + b_i) \tag{4.50}$$

The root mean square difference between the migrated features and the target features is used as the loss function, and the optimal migration parameters are obtained through optimization using the random gradient descent method. The random gradient descent method randomly uses part of the data in each iteration to find the optimal descent direction, which improves the efficiency of gradient operation compared to full gradient descent.

After obtaining a feature transfer model based on early data according to the above method, it is applied to all normalized source domain features. After obtaining transfer features that conform to the 0-1 distribution, the target domain anti-normalization function is applied to restore and obtain the final transfer features. However, due to the specificity of some features, not all features can be successfully transferred, so it is necessary to perform secondary selection on the migrated features. This selection is still based on pre wear data, which uses the first 150 sets of data between the migration features and the target features to calculate the maximum mean discrepancy (MMD) between the corresponding features, and selects the migration features that meet the MMD distance of less than 0.5, which is the successful migration feature. The maximum mean difference function can be used to calculate the distance between two feature distributions, as shown in Equation 4.51, where $X_{tr,i}$ is the transfer feature, $X_{t,i}$ are target features, and φ is the kernel function. The Gaussian kernel function is still used here.

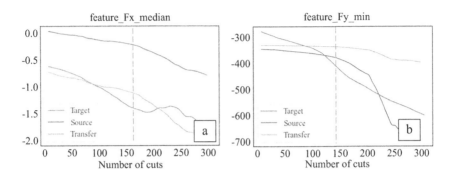

FIGURE 4.39 Example of feature transfer. a) Good case of transferred feature; b) Bad case of transferred feature.

$$MMD\left(X_{tr,i},\ X_{t,i}\right) = \left|\left|\frac{1}{n_{tr}}\sum_{j_{tr}=1}^{n_{tr}}\phi\left(X_{tr,ij}\right) - \frac{1}{n_t}\sum_{j_t=1}^{n_t}\phi\left(X_{t,ij_s}\right)\right|\right|_H^2 \quad (4.51)$$

Figure 4.39 shows the comparison of source features, migration features, and target features. It can be seen that the successfully migrated features are generally closer to the target features than the source features, but still maintain the distribution trend of the source features. Although the characteristics of migration failure are relatively close to the target features in local areas, the overall differences are significant.

After screening, the number of successfully transferred signal features from Experiment 1 to Experiment 2 was 58, and the number of successfully transferred signal features from Experiment 1 to Experiment 3 was 38. This proves that this transfer method can effectively transfer some signal features, and the number of successfully transferred features is still sufficient for the establishment of wear prediction models.

4.1.4.4.2 Verification of migration results

Considering the strong predictive ability of the RF algorithm in multiple operating conditions, the wear prediction models in this section all use the RF algorithm, and its hyperparameter selection is shown in previous tables. In order to compare the improvement of feature transfer in predicting the accuracy of new operating conditions, four comparative models are established here, with the main difference being the different training datasets:

Model 1: The training set is an unmigrated and unfiltered source dataset, containing 192 features.
Model 2: The training set is a source dataset that has not been transferred but has undergone initial filtering of features, containing 131 features.

TABLE 4.8

Results Evaluation of Transfer Learning

Test Dataset	Model	R^2	MSE	RMSE
Dataset 2	Model 1	0.4394	4.6935	22.0285
	Model 2	0.4683	4.7206	22.2842
	Model 3	0.8400	2.7537	7.5827
	Model 4	0.9617	0.5663	0.3207
Dataset 3	Model 1	0.3561	0.6919	0.4787
	Model 2	0.2929	0.6692	0.4478
	Model 3	0.6353	1.1308	1.2787
	Model 4	0.9699	0.3476	0.1208

Model 3: The training set is a transfer dataset with 131 features that have not undergone secondary filtering after transfer features.

Model 4: The training set is a migrated dataset that has undergone secondary filtering based on MMD distance. For the scene with the target set being Experiment 2, it contains 58 features. For the scenario with the target set of Experiment 3, it contains 38 features.

Table 4.8 presents the evaluation of the prediction results of the four models mentioned above on two test datasets. It can be seen that the prediction accuracy of the migrated model is significantly improved compared to before migration, and filtering the failed migration features can further improve the prediction accuracy. Finally, the decision coefficient R^2 of the migration prediction model can be increased from around 0.4 to 0.96 for the prediction results under new operating conditions, demonstrating the effectiveness of the feature migration method for predicting milling cutter wear under multiple operating conditions.

Figure 4.40 shows the comparison between the predicted wear values of the four models and the true values in the curve domain. It can be seen that Model 4

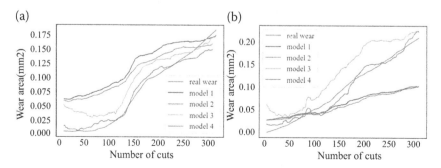

FIGURE 4.40 Comparison of predict wear and real wear for Model 4. a) The test set is dataset 2; b) The test set is dataset 3.

established after feature migration and filtering is the closest to the true wear values, and its accuracy is much higher than that of the model established without feature migration.

4.1.4.5 Summary

This section first prepares the dataset. Based on the feature extraction dataset from the previous section, filtering and data amplification were performed. Then, the correlation coefficients between each signal feature and wear were calculated, and preliminary feature screening was conducted. After obtaining the dataset, the accuracy and computational efficiency of three wear prediction algorithms, namely SVM algorithm, RF algorithm, and BPNN algorithm, were compared in a single operating scenario. The three methods all have high accuracy in single operating conditions, and the BPNN is slightly slower in calculation time. However, for multi-operating conditions, the model trained in a certain operating condition is directly applied to new operating condition data, resulting in lower prediction accuracy. Therefore, a feature transfer method was proposed in a multi-operating scenario, and the accuracy of the prediction model established by the successfully transferred features was significantly improved. The decision coefficient was increased from about 0.4 to about 0.96, verifying the effectiveness of this method.

4.1.5 SUMMARY OF THE CASE

Cutting tools are the core component of machine tools, and their wear degree is closely related to machining quality. Therefore, predicting tool wear is crucial. This case study focuses on coated surface milling cutters, extracting the time domain and frequency domain features of cutting force signals, vibration signals, and acoustic emission signals. Real wear is measured in place using visual methods as label values, and different prediction algorithms are compared to establish a wear prediction model. At the same time, the feature transfer method is applied to transfer the signal features and wear features extracted from historical tool data to the establishment of a prediction model for the target tool under new working conditions. This solves the problem of insufficient samples in new working conditions and improves the accuracy and universality of the tool wear prediction model. The main work and conclusions are as follows:

1. Completed the design of the tool wear experimental plan and the construction of the experimental platform, including the construction of the experimental platform, the selection of workpiece materials and processing parameters, signal acquisition methods, and experimental steps. Seven sets of wear experiments were conducted on coated surface milling cutters, and preliminary analysis was conducted on the experimental results. The experimental results show that the form of tool wear is greatly influenced by the hardness of the workpiece material, and there are two types of wear of surface milling cutters:

coating detachment and pits, especially for machining harder materials. The wear images obtained from this experiment, as well as signal data such as cutting force, torque, vibration, and acoustic emission, lay the foundation for subsequent feature extraction and wear prediction.

2. Proposed a machine vision-based in-situ wear measurement method for obtaining standard wear values when establishing prediction models. For tool wear images, preprocessing such as image cropping, graying, and filtering noise reduction are first performed. Then, after comparing some edge recognition and region recognition algorithms, it is confirmed to use the edge recognition method based on the Canny algorithm, and the recognition results are improved to obtain the wear areas corresponding to the two wear forms of pits and coating detachment. Finally, the weighted method is used to obtain the maximum wear width of the secondary and rear tool surfaces, respectively. The average wear width, wear area, and other wear characteristics facilitate the establishment of subsequent prediction models.

3. Based on the force signals, torque signals, vibration signals, and acoustic emission signals obtained from multiple sensors, a total of 12 statistical features were extracted using time domain and frequency domain analysis methods, including maximum, minimum, mean, median, standard deviation, root mean square, skewness, kurtosis, waveform factor, peak factor, pulse factor, and margin factor. Finally, 192 signal features were obtained. Then, based on the Pearson correlation coefficient, a preliminary feature screening was conducted to eliminate features with weaker correlation with wear, leaving 131 signal features after screening.

4. A single-condition wear prediction model was established by considering three sets of datasets with the same workpiece material and different working conditions. The accuracy and computational efficiency of three wear prediction algorithms, namely SVM algorithm, RF algorithm, and BPNN algorithm, were compared. The results show that all three methods have high accuracy in single operating conditions, with a decision coefficient R^2 of over 0.999. However, in terms of computational time, the training time of the BPNN is relatively long.

5. In response to the problem of low accuracy of prediction results when existing models are directly applied to new operating conditions data in multi-operating conditions scenarios, a linear feature transfer method is proposed by combining historical operating conditions with pre wear data of new operating conditions. The successfully transferred features are used to establish a new prediction model using the RF algorithm. The feasibility of this method was verified using the data from Experiment 1 as the source dataset, and the data from Experiment 2 and Experiment 3 as the target dataset, respectively. The results showed

that the accuracy of the established prediction model was significantly improved, and the determination coefficient was increased from about 0.4 to about 0.96, demonstrating the effectiveness of this method.

4.2 VISUAL MONITORING OF WORN SURFACE MORPHOLOGY

Wear is widely present in the industrial field and is the main cause of mechanical equipment failure or failure. Realizing real-time in-situ monitoring of material wear status helps to explore the evolution of material wear, providing support for mechanical equipment fault warning, life assessment, and operation maintenance. The morphology of worn surfaces can intuitively reflect the wear status of materials and reveal their wear mechanisms. Currently, contact and non-contact analysis techniques for worn surface morphology cannot achieve real-time in-situ monitoring of wear morphology. Meanwhile, relevant studies have shown that there is a certain correlation between the derived information such as force, vibration, sound, and sound pressure during the friction process and the morphology of the worn surface. This chapter explores a multi-source information fusion method for monitoring the wear status of test pieces in basic friction and wear tests, revealing the correlation between the wear surface morphology and multi-source information such as force, vibration, sound, and sound pressure during the friction test process, thereby achieving the reconstruction of wear surface morphology. The main research work of this case study is as follows:

1. A multi-source information collection system for friction process has been constructed, achieving the collection of multi-source information and wear surface morphology during friction testing. First, the environmental design theory was applied to design a multi-source information collection system for the friction process. A multi-source information collection system was built based on a multifunctional friction and wear testing machine. At the same time, a basic friction and wear test plan was designed under multiple working conditions, achieving the collection of vibration, sound, sound pressure, and other signals during the friction test process, as well as the acquisition of wear surface morphology. This provides a raw dataset for subsequent multi-source information fusion research.

2. Extracted multi-source information of friction process and key features of worn surface morphology, providing theoretical support for information correlation. For the multi-source information of the friction process collected in the basic friction test, digital signal processing methods are used to extract universal time and frequency domain features, while proprietary features are extracted for special acoustic signals. Extract features characterizing the roughness and periodicity of the worn surface morphology obtained in the experiment, providing a data basis for reconstructing the worn surface morphology.

3. A wear surface morphology feature monitoring method based on multi-source information was proposed, and combined with rough surface simulation method, the reconstruction of wear surface morphology was achieved. The correlation between wear surface morphology features and multi-source information features of friction processes was examined through correlation analysis. Different machine learning methods were used to quantify the correlation between the two. Based on this, a reconstruction method for wear surface morphology was proposed by combining rough surface simulation methods. The feasibility of this method was verified through experimental data.

4.2.1 RESEARCH BACKGROUND OF VISUAL MONITORING OF WORN SURFACE MORPHOLOGY

The issue of wear and tear widely exists in the industrial field (Figure 4.41), which is an important factor affecting the stability of mechanical equipment and also one of the key factors causing many mechanical equipment failures. The issue of wear and tear has gradually put forward more urgent requirements for the anti-friction and wear design and wear status monitoring of equipment operating in special working conditions such as space environment, high temperature, low temperature, high-speed, high humidity, and corrosive environment in the industrial field.

In the aerospace field, the contact surfaces of internal connections such as structural components and engine components of aircraft may experience slight relative motion, leading to fretting wear issues, leading to component failure or failure, which can affect the flight stability of the aircraft. In addition, key components used in aviation engines to improve flight stability may malfunction during operation due to wear and tear of key components, such as the lining of the adjustable static blade adjustment mechanism being affected by extreme working conditions such as high temperature and heavy load. The failure of key components in the aerospace field may lead to serious flight accidents and aviation safety issues. In order to achieve early warning of key component failures in the aerospace field, it is necessary to monitor the wear status of key components and clarify the wear evolution law of key components.

In the field of railway transportation, wheel rail contact, as an open contact system, is influenced by the synergistic effects of temperature, humidity, and speed, among which environmental conditions have a significant impact on wheel rail wear. Wheel rail wear can reduce passenger comfort and increase railway operation and maintenance costs. If the wear is severe, it is necessary to replace tracks or wheels that cannot continue to serve. If severely worn wheels and rails are not detected and replaced in a timely manner, it may cause serious railway traffic accidents and trigger a series of adverse effects. In order to achieve early warning of wheel rail failure, it is necessary to seek a wheel rail wear monitoring technology to obtain real-time wear status of the wheel rail.

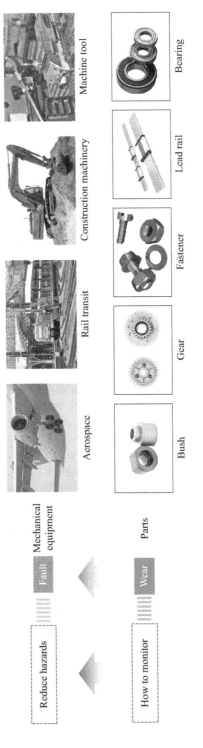

FIGURE 4.41 Universality of wear.

In the field of manufacturing and processing, tool wear is an important issue. As a key component of machining, cutting tools inevitably wear and tear against the workpiece during the machining process. When tool wear accumulates to a certain extent, the vibration intensification and temperature rise during the machining process will have a significant impact on the machining quality of the workpiece, and the machining speed will also be correspondingly slowed down, leading to the waste of machining resources and time costs, and even having a negative impact on the machine tool itself. Therefore, it is necessary to monitor the real-time wear of the tool to predict its remaining service life and replace the tool before excessive wear in order to improve production efficiency and product quality.

In order to analyze the friction and wear performance of materials and reveal the effects of different factors on material properties, domestic and foreign researchers have conducted a large number of friction and wear tests for different research objects, including basic friction and wear tests and simulated sample tests. Among them, the basic friction and wear test is carried out on a universal friction and wear testing machine based on specific operating conditions. The sample motion forms adopted include rotational motion and reciprocating motion, and the contact forms include surface contact, line contact, and point contact. The specific motion and contact forms selected depend on the specific research object. Basic friction and wear tests are usually used to study the effects of different friction pair materials, lubricants, and test conditions on the friction and wear performance of materials. The simulated sample test is mainly conducted on a friction and wear testing machine developed by the laboratory, using simplified components that are close to reality to replace the friction pairs in the basic friction and wear test, and conducting the test under simulated working conditions. Whether it is a basic friction and wear test or a simulated sample test, the material wear rate, wear amount, friction coefficient changes during the test process, and wear surface morphology are the focus of researchers' attention. Among them, wear rate, wear amount, and wear surface morphology are the main indicators to describe the friction and wear resistance performance of materials. However, in these experiments, only the changes in sample quality before and after the experiment, as well as the surface morphology of the sample, can be directly obtained, which makes it difficult to monitor and analyze the wear evolution in real time during the experiment process.

In the field of manufacturing and processing, research on monitoring technology for tool wear has become relatively mature. Researchers achieve monitoring of tool wear by collecting multi-source sensor signals in real time during the machining process. The commonly used signals for tool wear monitoring include force signals, vibration signals, acoustic emission signals, spindle current signals, etc. In basic friction and wear tests, tribological system information can be roughly divided into three categories, including input information, system environment information, and output information. Among them, the output information can be further divided into (1) information that directly reflects the friction and wear performance of materials such as friction coefficient, wear amount, and wear surface morphology, and (2) friction test by-products such as sound and vibration. In this study, the information generated during the friction process is collectively referred to as tribological information. In traditional tribological experimental research, researchers mainly focus on the correlation between information such as friction coefficient, wear amount, and wear surface morphology with input parameters. This ignores the reflection of by-product information on friction and wear phenomena during the friction and wear test process, and also overlooks the wear evolution law of materials during the friction and wear process.

In summary, the issue of wear introduces uncertain factors into the stable operation of mechanical equipment in the industrial field. By combining information technology with basic tribology research, the correlation between wear status and friction by-product information can be achieved, providing a technical basis for monitoring the wear status of mechanical equipment.

In order to study the evolution law of wear state in the process of material friction and wear, this case introduces information technology into tribological experimental research. By integrating and analyzing derived information such as force, vibration, sound, and sound pressure during the friction and wear process, real-time monitoring of the surface wear state of friction pairs can be achieved.

Scientific research significance: In the past, due to limitations in conditions, researchers could only speculate and analyze possible phenomena during material friction and wear by comparing the quality and surface morphology changes of samples before and after wear. The use of multi-source sensor information in the friction and wear process to predict material wear status has broken through the challenge of real-time in-situ monitoring, helping researchers explore the evolution law of material wear status during the friction and wear process, and providing a basis for accurately evaluating the friction and wear performance of materials.

Engineering application value: Real-time in-situ monitoring of material wear status during friction and wear provides a technical foundation for monitoring the operating status of key components in mechanical equipment. With the further deepening of research, this method can be extended from basic friction and wear tests to actual service components, providing a solid foundation for fault warning and life assessment of components, and also providing an algorithmic basis for the development of wear surface morphology analysis instruments.

4.2.2 Design and Data Acquisition of Multi-Source Tribology Information Collection System

4.2.2.1 Introduction

The design and development of a friction process information collection system can provide a data foundation for multi-source information fusion and multi-disciplinary information association mapping. The existing friction process information collection systems can be roughly divided into two types. One type is the information collection system that comes with universal friction and wear testing machines, which usually collects relatively single information and is difficult to obtain multi-disciplinary friction process information. Another type of information collection system is usually built into a self-developed friction and wear test bench in the laboratory, lacking a certain degree of universality. Therefore, in order to more comprehensively and accurately monitor the friction and wear status and evolution law of materials, it is necessary to design and build a multi-source information collection system for the friction process based on a universal friction and wear testing machine. This chapter uses environmental design theory to design and develop a multi-source information collection system for friction process based on a universal friction and wear testing machine. Real-time and in-situ collection of friction and wear test data is achieved through conducting pin disk friction and wear tests. At the same time, the surface morphology of the sample disk during the friction process was collected using a white light interferometer using a single sample with multiple starts and stops, providing a basis for subsequent data processing.

4.2.2.2 Development of a multi-source information collection system based on environmental design theory

In order to collect multi-source information during the basic friction and wear testing process, this chapter uses the Rtec-5000S multifunctional friction and wear testing machine as the carrier to design and develop a multi-source information collection system for the friction process. Due to the development of a multi-source information collection system on a highly versatile testing machine, it is necessary to consider the impact of environmental factors on system design. Environmental based design (EBD) can comprehensively consider the environmental impact during product design and development, guide researchers to identify conflicts in product design, and generate corresponding solutions, thereby completing product design and development. Therefore, this chapter is based on environmental design theory to design a multi-source information acquisition system for friction processes and develop a friction process information acquisition system that includes signal acquisition functions such as vibration signals, sound pressure signals, and sound signals, providing support for the collection of multi-source information during friction testing. On this basis, conduct friction and wear tests on the pin plate foundation, design the test process and plan using a single sample with multiple starts and stops, collect multi-source information on the friction process and wear surface morphology during the test process, and provide data sources for subsequent data processing.

4.2.2.2.1 Environmental design theory

Environmental design theory is a design methodology derived from logical steps based on the axiomatic theory of design modeling. Under this theoretical framework, design problems are implicit in the environment in which the product is located, including three parts: the expected working environment of the product, the environmental requirements for product structure, and the environmental requirements for product performance. Among them, the environment includes natural environment, artificial environment, and human factor environment. In the theoretical model of environmental design, designers take the environment in which the product is located as the object, explore the implicit product requirements, and gradually improve the product design plan through conceptual design and other methods. In this process, designers need to accurately and comprehensively identify hidden requirements and conditions in the environment, in order to form a comprehensive solution and provide corresponding services for the environment.

In the process of product design based on environmental design theory, there are three core steps involved: environmental analysis, conflict identification, and solution generation, as shown in Figure 4.42. Designers need to constantly repeat and iterate these steps to generate the final solution. Among them, environmental analysis aims to identify the various components that make up the expected working environment of the product and the relationships between them. By comprehensively considering customer needs and other potential needs, designers can build a relatively complete environmental system. The analysis of environmental systems can be achieved through the recursive object model (ROM). After completing the construction of the ROM diagram, designers can clarify the conflicts between different objects contained in the environment. After generating a corresponding solution for a certain conflict, updating the ROM diagram and repeating the previous steps can gradually eliminate the conflicts in the environment and obtain the final solution. In the application process of environmental design theory, the identification of each key conflict provides a fundamental driving force for the update of the solution. When there are no significant conflicts in the environmental system, it means that the current solution can meet all customer needs and other implicit needs. Therefore, the conflict recognition step plays an important role in the use of environmental design theory.

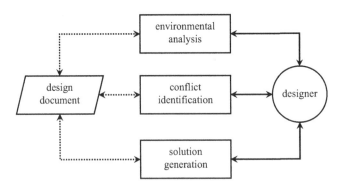

FIGURE 4.42 Application process of environment-based design methodology.

4.2.2.2.2 Design model and conflict analysis

The friction process multi-source information collection system developed in this chapter is built using a multifunctional friction and wear testing machine as the carrier, and is applied in the field of basic tribology testing. In the environmental analysis stage, designers need to clarify the current operational process of basic friction and wear tests and the processing of test data, in order to sort out the structure and performance requirements of the friction process multi-source information collection system. After sorting out, corresponding solutions are proposed for key conflict points in each iteration based on the constructed ROM diagram, and the design and construction of the information collection system are ultimately completed. The detailed element definitions of the ROM are shown in Table 4.9.

In response to the design goal of "designing an information collection system to obtain multi-source sensor information in basic friction tests to evaluate the friction and wear status of materials," an ROM diagram is established, as shown in Figure 4.43. Among them, the three objects of "system," "information," and "state" are subject to the most constraints in the ROM diagram and can be used as key components of the environment in environmental design theory.

By questioning key points, the current ROM map can be further refined. For example, the answer to "which sensor information was obtained from the basic friction test" is "the obtained multi-source sensor information includes tribological state information and derived state information." By questioning key components in the environment, the correlation between key points can be further explored and clarified, such as how to evaluate the friction and wear status of materials. By conducting data mining and other operations on the obtained multi-source information of the friction process, the association

TABLE 4.9
Definition of ROM Parameters

Type		Graphical Representation	Description
Object	Object	O	Everything in this world can be an object
	Complex object	O	Objects that contain at least two other objects
Relationship	Binding relationship	●——→	A relationship between objects that describes, limits, or materializes
	Contact relationships	- - - ι - - →	Connecting two objects that are not bound to each other
	Predicate relationship	— ρ —→	Describing the behavior of one object with respect to another, or describing the state of an object

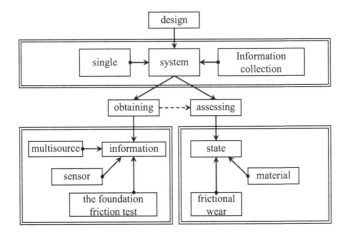

FIGURE 4.43 ROM of multi-source tribological information acquisition system.

between multi-source information and material friction and wear state parameters is constructed. By repeatedly questioning key points, the ROM graph is updated to Figure 4.44.

Based on the final ROM map obtained from Figure 4.4, key conflicts in the environment can be identified and analyzed, and solutions can be sought one by one to overcome each conflict. Table 4.2 lists the key conflicts that exist in design projects.

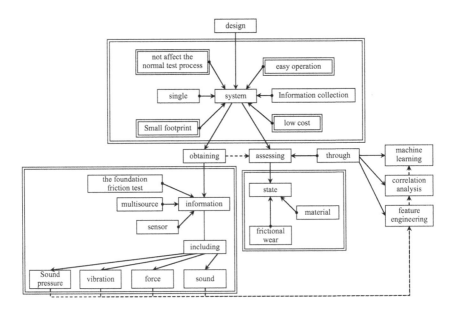

FIGURE 4.44 Updated ROM of multi-source tribological information acquisition system.

TABLE 4.10

Key Conflicts During the Design of the Multi-Source Tribological Information Acquisition System

Number	Conflicts	
1	Testers want to observe the operating state of the test and the wear state of the material in real time.	Wear is difficult to observe in-situ.
2	It is difficult to fully grasp the state of tribology with a single piece of information.	The existing friction and wear testing machines lack multi-source information acquisition function.

Corresponding solutions can be proposed to address the key conflicts in the design projects excavated in Table 4.10, thus achieving the development of a multi-source information collection system for friction processes.

4.2.2.2.3 Development of a multi-source information collection system

Tribological behavior has a multi-disciplinary coupling characteristic, and the friction process involves various phenomena, such as force, heat, electricity, light, and magnetism. Therefore, a single physical, chemical, material, and mechanical information set cannot accurately and completely describe the characteristics of tribological systems. Therefore, in order to build a relatively complete information unit for tribology systems, it is first necessary to consider from the perspective of information collection. In tribological experimental research, the information directly related to the friction and wear status mostly exists in an implicit form in the tribological system, such as wear amount, wear depth, wear rate, and wear surface morphology. They are closely related to the friction test conditions, as well as the friction material characteristics, and are important sources of information for analyzing and studying the behavior of tribological systems. The difficulty in obtaining this information makes it hard to directly monitor the friction and wear status. By utilizing friction information technology, the monitoring of friction and wear status can be achieved by establishing a correlation between easily obtainable tribological derivative quantities and difficult-to-obtain tribological state quantities.

To ensure the integrity of tribological information, this chapter has carried out the design of a multi-source information collection system for friction processes. As shown in Figure 4.45, the information collection system is designed and built based on the Rtec-5000S multifunctional friction and wear testing machine, consisting of different types of sensors, data collection cards, and computers equipped with software systems. Among them, sensors include acceleration sensors for collecting vibration signals, laser displacement sensors, sound pressure sensors for collecting sound pressure signals, and microphones for

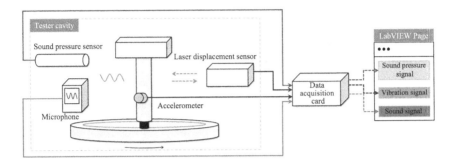

FIGURE 4.45   Scheme design of multi-source tribological information acquisition system.

collecting sound signals, which are deployed inside the chamber of the testing machine. By connecting to the data acquisition card, vibration signals and acoustic signals can be obtained and stored in real time on the software interface. In addition, due to the built-in information collection system of the friction and wear testing machine having the function of collecting force and displacement signals, these signals will also be integrated into the subsequent dataset construction. To obtain the morphology of the worn surface during the experiment, a white light interferometer equipped with the testing machine was used for photography.

The equipment and sensors used to obtain multi-source tribological information during the friction test process are detailed as follows:

1. Rtec-5000S multifunctional friction and wear testing machine
   The friction process multi-source information collection system designed in this chapter is deployed inside the cavity of the Rtec-5000S multifunctional friction and wear testing machine. The testing machine comes with an information collection system that can achieve the collection and processing of some basic data, as shown in Figure 4.46. The built-in information collection system of the testing machine can collect real-time information such as load, friction force, friction coefficient, and upper sample height during the testing process, and automatically store it as a CSV format file. Among them, the real-time friction coefficient is obtained by dividing the friction force measured by the built-in force sensor by the load. During the experiment, the load will fluctuate around the set value. The sampling frequency of the signal is 1000 Hz, and the system uses the sliding average method to downsample the signal to 100 Hz by default. In addition, the required worn surface morphology is collected by the built-in 3D optical profilometer of the testing machine, which has four optical imaging modes: confocal, white light interference, dark field, and bright field. This chapter mainly uses white light interference mode for morphology collection, with a resolution of 1.8117 in the X and Y directions μ M.

(b) Built-in 3D optical topographer

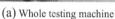
(a) Whole testing machine

(c) Built-in information collection system

FIGURE 4.46 Rtec-5000S multifunctional tribometer.

2. Laser displacement sensor

The KEYENCE LK-G80 laser displacement sensor is used to obtain the X-direction vibration information of the upper sample axis during the friction test process. The effective measurement range is 80 ± 2 mm, with an accuracy of 1 μ m. Sensitivity is 10 mV/ μ M. In the preparation stage of the experiment, fix it at a position about 80 mm to the right of the upper specimen axis, as shown on the right side of Figure 4.47(a). To record the data of the laser displacement sensor in real time, it is connected to the upper computer through a data acquisition card to achieve real-time collection, transmission, and storage of the X-direction vibration signal. Its sampling frequency is 1000 Hz.

3. Acceleration sensor

The two piezoelectric acceleration sensors in Figure 4.47(a) are used to collect vibration information in the Y and Z directions of the upper sample axis, with a sensitivity of approximately 10 mV/(m · s ^ (-2)). Among them, the Y-direction acceleration sensor is fixed on the upper sample axis using adhesive, and the Z-direction acceleration sensor is fixed on the bottom of the loading platform using magnetic attraction. To record the data of acceleration sensors in real time, two sensors are amplified through multi-channel constant current adapters, and then connected to the upper

(a) Inside of the tester cavity   (c) Multi-channel constant
                                             current adapter

FIGURE 4.47   Multi-source tribological information acquisition system.

computer through a data acquisition card to achieve real-time acquisition, transmission, and storage of acceleration signals.

4. Sound pressure sensor

The GRAS-46AD sound pressure sensor is used to collect the sound pressure signal inside the testing machine, with a sampling frequency of 1000 Hz and a sensitivity of 50 mV/Pa. In the preparation stage of the experiment, fix it diagonally above the contact area of the pin and disc, as shown on the left side of Figure 4.47(a). To record the data of the sound pressure sensor in real time, it is also amplified through a multi-channel constant current adapter, and then connected to the upper computer through a data acquisition card to achieve real-time collection, transmission, and storage of the sound pressure signal.

5. Digital microphone

The DR-05X digital microphone using TASCAM collects sound signals during the testing process, and can ultimately generate files in wave format. In the preparation stage of the experiment, the digital microphone is placed on the platform inside the testing machine and close to the contact position of the pin plate, as shown in Figure 4.47(a). The sampling frequency of the digital microphone is 48000 Hz.

The data acquisition card and multi-channel constant current adapter in the information collection system are shown in Figure 4.47(b) and (c). Among them, the constant current adapter can support signal transmission and amplification of four channels, with × 1 and × there are 10 gain levels, and three of them are mainly used in this experiment.

4.2.2.3   Experimental plan design and data acquisition

This case selected Ti6Al4V pins as the upper sample and copper discs as the lower sample, designed and developed a test plan with load as the variable, and conducted corresponding pin disc basic friction and wear tests, achieving the acquisition of multi-source information and wear surface morphology during the friction test process, providing a data basis for subsequent signal processing.

4.2.2.3.1   Experimental scheme

The experiment used Ti6Al4V pin as the upper sample, with a diameter of 6.35 mm, and a copper plate as the lower sample, with a diameter of 69.85 mm and a thickness of 6.35 mm, as shown in Figure 4.48.

To verify the applicability of the multi-source information fusion method proposed in this chapter for different test conditions, this chapter designed and conducted basic friction and wear tests under different conditions, with load as the variable. First, in order to eliminate the impact of environmental noise on the experiment, a no-load test was designed as a control group. Secondly, in order to obtain multi-source information and wear surface morphology under different load conditions, experiments were designed under different load conditions. As each group of experiments was conducted in the form of multiple starts and stops of a single sample, each group of experiments was composed of several groups. Repeat each experiment three times to rule out randomness. The specific experimental plan is shown in Table 4.11.

4.2.2.3.2   Test process

The overall process of the experiment is shown in Figure 4.49, and the specific steps are as follows:

1. **Sample preparation:** New test samples are selected for each group of experiments. Before the experiment, Ti6Al4V is soaked in anhydrous ethanol and placed in an ultrasonic cleaning machine to remove surface

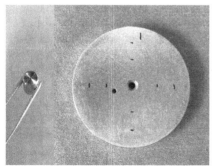 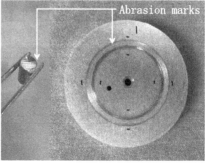

(a) Pins and discs before friction-wear test   (b) Pins and discs after friction-wear test

FIGURE 4.48   Test samples.

TABLE 4.11

Test Plan

Number	Load/N	Rotation Radius/mm	Rotate Speed/rpm	Test Duration/min	Number of Groups/ Group
0	none	20	60	3	1
1	5.0	20	60	3	30
2	7.5	20	60	3	30
3	10.0	20	60	3	30

Samples   Shooting area marker   Frictional test   Surface topography

FIGURE 4.49   Test procedures.

 oil stains. At the same time, the oil stains on the surface of the sample disk are wiped with anhydrous ethanol.

2. **White light shooting area marking:** Due to the movement mode adopted in the experiment being the rotation of the lower sample, the resulting grinding marks will form a circular ring. In order to accurately obtain the morphology of the worn surface during the test process, after the sample is prepared, mark with a marking pen at equal intervals at the radius of 15 and 25 mm on the sample disk, to facilitate the acquisition of morphology during the test process.

3. **Obtaining the initial surface morphology of the sample disk:** Fix the cleaned sample pin and the marked sample disk inside the testing machine with fixtures, move the sample disk below the three-dimensional optical topography instrument, and collect the morphology of the four marked positions using Rtec morphology acquisition software as the initial surface morphology of the sample disk.

4. **Friction test:** Use Rtec software to set the test conditions and adjust the position of the testing machine platform so that the pin is at the appropriate rotation radius. To collect experimental data, turn on the information collection software, digital microphone, and testing machine of the upper computer in sequence. During this process, keep the testing machine door closed. According to the experimental program setting, the relative motion of the upper sample pin and the

lower sample disk will begin after the measured load stabilizes at the set load for a certain period of time. At this point, the driving motor starts to rotate, driving the sample disk to rotate relative to the sample pin. When the test duration reaches the set value, the test is completed, and the testing machine, upper computer information acquisition software, and digital microphone are sequentially turned off to complete the data collection of the friction process for a group.

Move the sample disk to the bottom of the 3D optical morphometer again, and collect the morphology of the four marked positions using Rtec morphology acquisition software to obtain the worn surface morphology of the sample disk.

5. Repeat Steps 3 and 4 until the number of test groups reaches 30. Complete the collection and storage of the entire set of test data.

4.2.2.4   Summary

This section designs and develops a multi-source information collection system for friction process based on the Rtec-5000S multifunctional friction and wear testing machine, achieving the collection of vibration, sound, sound pressure, and other multi-source information during the friction testing process. First, the environmental design theory was applied to the field of tribology experimental research, and the scheme design and physical construction of a multi-source information collection system were achieved by comprehensively considering various key conflicts. On this basis, a pin disc friction and wear test plan was designed and developed under multiple working conditions. Considering its repeatability and universality, control experiments and control variable experiments were conducted, respectively, achieving the collection of multi-source process information and wear surface morphology during the friction test process, providing a data foundation for subsequent multi-source information fusion.

4.2.3   Feature Extraction of Tribological Information

4.2.3.1   Introduction

The multi-source information during friction and wear testing is susceptible to environmental interference, and due to high sampling frequency and large data volume, it is usually difficult to directly calculate and analyze as input to the model. Therefore, it is necessary to process and analyze the multi-source signals and wear surface morphology collected accordingly. This chapter first pre-processes the collected vibration signals, sound pressure signals, and sound signals to eliminate the influence of environmental factors. Then, feature engineering methods are used to extract features that reflect the friction and wear status of materials from multi-source information. At the same time, statistical methods are used to extract features of wear surface morphology, providing a basis for the correlation analysis between multi-source information in the friction process and wear surface morphology.

4.2.3.2 Feature extraction of multi-source tribological information

Due to the susceptibility of multi-source information collected in friction and wear tests to environmental interference, this section adopts signal preprocessing methods such as data filtering and data cropping to reduce the impact of environmental interference on the test data. Meanwhile, due to the fact that the multi-source information of the friction process collected from the experiment contains multiple types of signals, in addition to extracting common time-frequency statistical features of the signal, key features were extracted for specific sound pressure and sound signals, providing input for the correlation between multi-source signals and wear surface morphology.

4.2.3.2.1 Signal preprocessing

Signal preprocessing mainly consists of two stages: signal filtering and signal clipping. Among them, signal filtering is used to eliminate interference factors during the experimental process, and signal clipping is used to select experimental process data from a period of time before each morphology shot. Here, the process of signal preprocessing is illustrated using experimental data with a load of 5 N as an example.

4.2.3.2.1.1 Signal filtering Figure 4.50(a) shows the data of the 28th group in the experiment. From top to bottom, they are the displacement signal in the X-direction, the acceleration signal in the Y-direction, the acceleration signal in the Z-direction, and the sound pressure and sound signal in the experiment. From

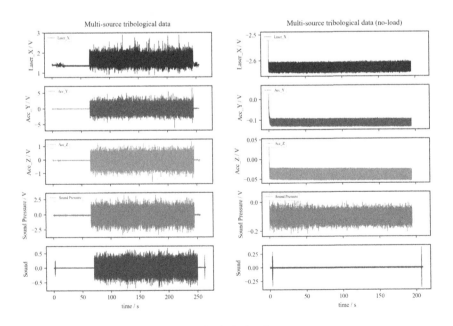

FIGURE 4.50   Raw multi-source tribological signals.

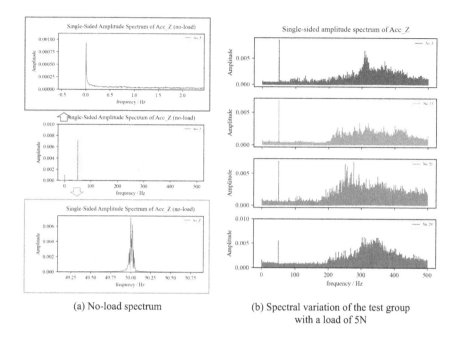

(a) No-load spectrum

(b) Spectral variation of the test group
with a load of 5N

FIGURE 4.51 Fourier spectrum of the acceleration signals in the Z-direction of the upper sample.

Figure 4.50(a), it can be seen that the signal values before and after the relative motion of the friction pair are not constant, resulting in lower amplitude environmental noise. In addition, there is a signal mutation before and after the start and end of the sound signal, which corresponds to the closing and opening of the testing machine door in the testing process. The phenomenon of noise signals can also be seen from the no-load test data shown in Figure 4.50(b). The use of a Butterworth filter can effectively eliminate the impact of noise in the environment. Therefore, this chapter uses a Butterworth filter to filter the signal, thereby retaining only the signal related to friction pair wear.

Taking the acceleration signal in the Z-direction of the sample axis (hereafter referred to as Acc_Z signal) as an example, in order to determine the specific filtering method and parameters, Fourier transform is first performed on it. From the no-load signal spectrum in Figure 4.51(a), it can be seen that there is stable low-frequency noise and 50 Hz power frequency interference. At the same time, as shown in Figure 4.51(b), the Fourier spectrum during the formal test process shows that the frequency band data above 200 Hz varies with the progress of the test, indicating a certain correlation between the frequency band data above 200 Hz and the wear status of the friction pair. Therefore, a Butterworth high-pass filter can be used to eliminate low-frequency noise interference and preserve the frequency band related to signal and material wear, with a specific filtering cutoff frequency set at 200 Hz. Figure 4.52 shows the Acc of some groups before

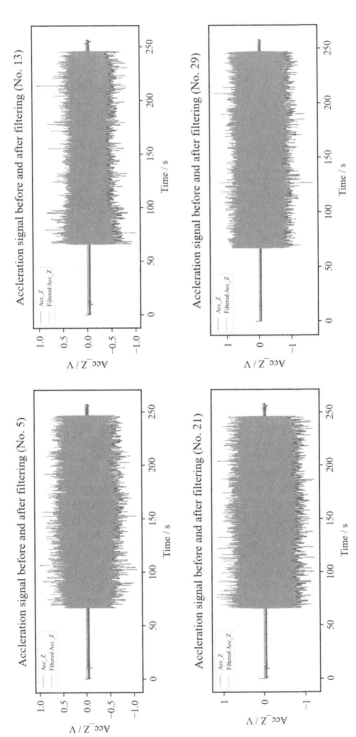

FIGURE 4.52 Acceleration signals in the Z-direction of the upper sample before and after filtering.

TABLE 4.12

Filtering Methods of Different Tribological Signals

Friction Process Signals	Filtering Forms	Cutoff Frequency (Low Pass)/Hz	Cutoff Frequency (High Pass)/Hz	Cutoff Frequency (Low Resistance)/Hz	Cutoff Frequency (High Resistance)/Hz
Upsample axis X-direction displacement	Bandpass filtering	400	70	/	/
Upsample axis Y-direction acceleration	High-pass filtering	/	70	/	/
Upsample axis Z-direction acceleration	High-pass filtering	/	200	/	/
Sound pressure	High-pass filtering	/	100	/	/
Sound	Band-resistance filtering	/	/	5000, 15000	10000, 20000

and after filtering Acc_Z signal, where the blue line represents the signal curve before filtering, and the yellow line represents the signal curve after filtering. From the figure, it can be seen that the amplitude of the signal curve before and after the relative motion of the friction pair decreases and fluctuates around 0, indicating that high-pass filtering can effectively filter out environmental noise interference during the friction test process.

Similarly, the Butterworth filter is used to filter other raw friction process signals. The specific filtering methods and parameters are shown in Table 4.12. Due to the low sampling frequency of the signal collected by the testing machine information collection system, the testing machine signal was not filtered.

4.2.3.2.1.2 Signal clipping To ensure the integrity of the test dataset, the multi-source information collection software is opened before the formal start of the friction test and closed after the friction test is completed. Meanwhile, from Figure 4.50(a), it can also be seen that the total time for the sensor to collect signals includes the time before and after the relative motion of the friction pair begins. Therefore, in order to obtain the sensor signal with an effective time length, it is necessary to crop the filtered signal. Taking Acc_Z signal as an example, since the signal amplitude during the friction test is significantly greater than the signal amplitude before the start of the test, a preset threshold can be used to determine the effective start and end time of the signal. That is, when the signal amplitude first exceeds the preset threshold, it is considered that the relative motion between the friction pairs starts at the current time. When the signal amplitude falls below the preset threshold for the first time, it is considered that the relative motion between the friction pairs has stopped. On this basis, in order to establish the correlation between signals and worn surface morphology, this chapter selects signals with a duration of 60 seconds before the actual end of each group of experiments as the research object. Figure 4.53 shows a portion of Acc_Z schematic diagram of the clipping method for the signal, where the blue curve represents the filtered Acc_Z signal curve, with the yellow transparent area representing the effective signal interval obtained after clipping, and the gray transparent area representing the signal interval of the trimmed part. Due to the high sampling frequency and large data volume of the cropped signal, it is necessary to extract and select features from the effective signal obtained through cropping.

4.2.3.2.2   General signal feature extraction

Common signal feature extraction methods can be roughly divided into two categories: time domain feature extraction and frequency domain feature extraction. Time domain feature extraction refers to the calculation of signal features from a statistical perspective based on time domain signals. Frequency domain feature extraction first obtains the spectrum of the signal through Fourier transform and then extracts useful statistical information from the spectrum.

Figure 4.53 shows schematic diagrams of trimming the acceleration signals in the Z-direction of the upper sample.

Time domain features can be divided into dimensional features and dimensionless features. Among them, the dimensional features include the maximum, minimum, range, mean, median, mode, standard deviation, root mean square value, mean square value, etc. of the signal. Dimensionless features include skewness, kurtosis, peak factor, waveform factor, pulse factor, margin factor, etc. The maximum, minimum, and range of signals are commonly used to determine impact type faults, while the root mean square value represents the effective value of the signal. In the field of fault diagnosis, it is commonly used to determine wear type faults; kurtosis is sensitive to impact signals and is commonly used to determine pitting loss faults. Peak factor is commonly used to characterize signal impact. The specific calculation method for the time domain features used in this chapter is shown in Table 4.13.

The frequency domain features used in this case also include maximum, mean, variance, mean square, skewness, and kurtosis. In addition, they also include frequency domain features unique to the spectrum such as center of gravity frequency, root mean square frequency, skewness frequency, and kurtosis frequency. Among them, the spectral mean represents the average energy of the signal, while the variance, skewness, and kurtosis of the spectrum represent the degree of concentration of the signal spectrum. The specific calculation method for frequency domain features is shown in Table 4.14.

After performing preprocessing operations such as signal filtering and signal clipping on the signals collected in the experiment, the time frequency and frequency domain features of the signals were extracted using the feature

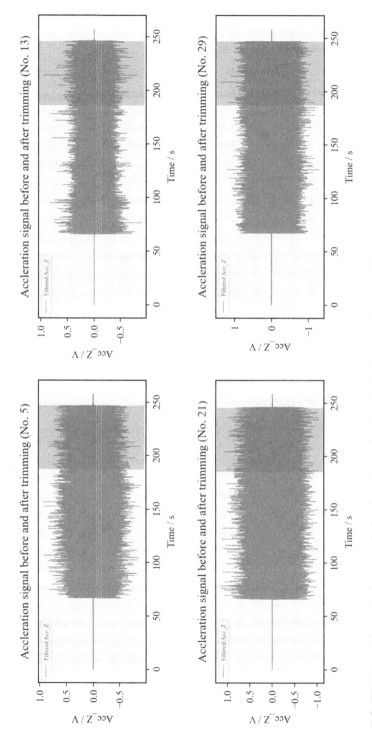

FIGURE 4.53 Schematic diagrams of trimming the acceleration signals in the Z-direction of the upper sample.

TABLE 4.13

General Time Domain Features of Signals

Features	Equations	Features	Equations
Maximum value	$\max(x_i)$	Quadratic mean	$\frac{1}{n}\sum_{i=0}^{n-1} x_i^2$
Minimum value	$\min(x_i)$	Skewness	$E\left[\left(\frac{x_i-\mu}{\sigma}\right)^3\right]$
Range	$\max(x_i) - \min(x_i)$	Kurtosis	$E\left[\left(\frac{x_i-\mu}{\sigma}\right)^4\right]$
Average value	$\mu = \frac{1}{n}\sum_{i=0}^{n-1} x_i$	Peaking factor	$\frac{\max(|x_i|)}{x_{rms}}$
Median	$\text{median}(x_i)$	Corrugation factor	$\frac{x_{rms}}{\frac{1}{n}\sum_{i=1}^{n-1}|x_i|}$
Mode	$\text{mode}(x_i)$	Impulse factor	$\frac{\max(|x_i|)}{\frac{1}{n}\sum_{i=0}^{n-1}|x_i|}$
Standard deviation	$\sigma = \sqrt{\frac{1}{n}\sum_{i=0}^{n-1}(x_i - \bar{x})^2}$	Margin factor	$\frac{\max(|x_i|)}{\left(\frac{1}{n}\sum_{i=0}^{n-1}\sqrt{|x_i|}\right)^2}$
Root mean square	$x_{rms} = \sqrt{\frac{1}{n}\sum_{i=0}^{n-1} x_i^2}$	/	/

TABLE 4.14

General Frequency Domain Features of Signals

Features	Equations	Features	Equations
Maximum value	$Z_1 = \max(F_k)$	Kurtosis	$Z_6 = E\left[\left(\frac{F_k-Z_2}{\sqrt{Z_3}}\right)^4\right]$
Average value	$Z_2 = \frac{1}{K}\sum_{k=0}^{K-1} F_k$	Gravity frequency	$Z_7 = \frac{\sum_{k=0}^{K-1}[f(k)\cdot F(k)]}{\sum_{k=0}^{K-1} F(k)}$
Variance	$Z_3 = \frac{1}{N-1}\sum_{k=0}^{K-1}(F_k - Z_2)^2$	RMS frequency	$Z_8 = \sqrt{\frac{\sum_{k=0}^{K-1}[(f(k)-Z_7)^2 F(k)]}{\sum_{k=0}^{K-1} F(k)}}$
Quadratic mean	$Z_4 = \frac{1}{K}\sum_{k=0}^{K-1} F_k^2$	Skewness frequency	$Z_9 = \frac{\sum_{k=0}^{K-1}[(f(k)-Z_7)^3 F(k)]}{\sum_{k=0}^{K-1} F(k)}$
Skewness	$Z_5 = E\left[\left(\frac{F_k-Z_2}{\sqrt{Z_3}}\right)^3\right]$	Kurtosis frequency	$Z_{10} = \frac{\sum_{k=0}^{K-1}[(f(k)-Z_7)^4 F(k)]}{\sum_{k=0}^{K-1} F(k)}$

calculation equation in Tables 4.13 and 4.14, respectively. In the process of extracting frequency domain features of acoustic signals, due to the presence of two frequency bands in the Fourier spectrum of sound pressure and sound signals, a Butterworth filter is used to separate the sound pressure and sound signals into two frequency bands for frequency domain feature extraction.

In Figure 4.54, under $5N$ load condition, the Acc_Z signal is used as an example to demonstrate the extracted universal time and frequency domain

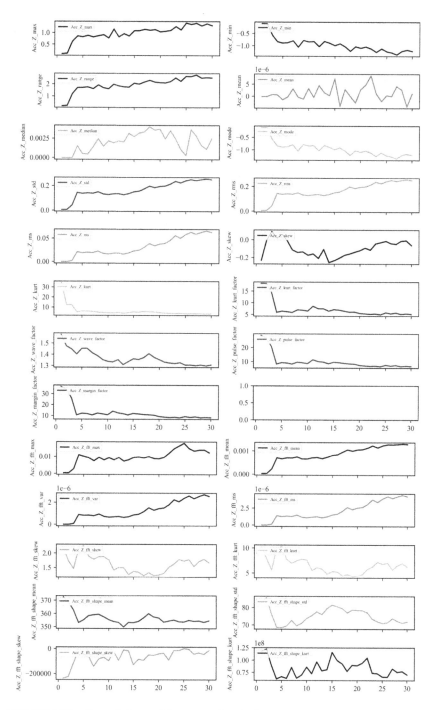

FIGURE 4.54 General time and frequency domain features of the acceleration signals in the Z-direction of the upper sample.

features. As shown in the figure, the partial temporal and frequency domain characteristics of the Acc_Z signal exhibit a certain trend, gradually increasing or decreasing with the evolution of material friction and wear conditions. At the same time, there are differences in magnitude between different features. In order to eliminate the influence of feature scale in subsequent information fusion, it is necessary to consider standardizing or normalizing the features.

4.2.3.2.3 Sound pressure signals feature extraction

Sound pressure signal is a type of acoustic signal, which is a change in pressure value caused by the disturbance of sound waves on atmospheric pressure. In the friction process multi-source information collection system built in this chapter, it is collected by sound pressure sensors. In the study of acoustic signals, the concept of sound pressure level (SPL) is commonly used to describe sound pressure with significant linear variations. The calculation process is as follows:

1. Calculate the root mean square value of sound pressure over time length T using the following Equation 4.52:

$$\bar{p} = \sqrt{\frac{1}{T} \int_0^T [p_{var}(t)]^2 dt} \tag{4.52}$$

   Among them, $p_{var}(t)$ represents the instantaneous amplitude of sound pressure.

2. Convert the root mean square value of sound pressure to the logarithmic scale using the following Equation 4.53 to obtain the SPL:

$$SPL(dB) = 20lg\left(\frac{\bar{p}}{p_{ref}}\right) = 10lg\left(\frac{\bar{p}^2}{p_{ref}^2}\right) \tag{4.53}$$

Among them, $p_{ref} = 2 \times 10^{-5}$Pa is the reference sound pressure value in the air.

The SPL changes during the test process under three operating conditions calculated using Equations (4.52) and (4.53) are shown in Figure 4.55. As shown in the figure, as the experiment progresses, the SPL characteristics show significant changes in the early stage and stable changes in the later stage, which is consistent with the evolution process of friction and wear states, namely, the initial stage of rapid wear and then the stage of stable wear and tear.

4.2.3.2.4 Audio signal feature extraction

Audio signal is another acoustic signal studied in this chapter, which is collected by a digital microphone. In the study of audio signals, Mel-scale frequency

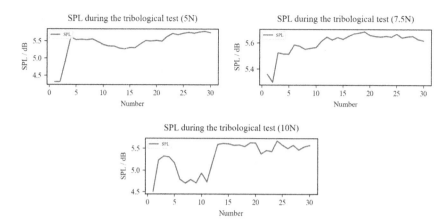

FIGURE 4.55 Evolution of sound pressure level under different load conditions.

cepstral coefficients (MFCC), chromaticity maps, spectral centroids, spectral bandwidth, spectral attenuation, and zero crossing rate are important information that researchers pay attention to, and are commonly used in fields such as music information retrieval and speech recognition classification.

MFCC is used to analyze signal features by simulating human hearing, and the solving process is relatively complex. The specific calculation steps are as follows:

1. **Preprocessing:** Use a pre-emphasis filter to amplify the high-frequency part of the audio signal, making the overall spectrum of the signal tend to balance, while also improving the signal-to-noise ratio. Then, divide the signal into frames to ensure better stability of the analyzed signal. Add windows to each frame of the signal, such as a triangular overlapping window or hamming window, to increase the coherence between adjacent frames.

2. **Calculate power spectrum:** Perform Fourier transform (STFT) on the preprocessed signal frames to obtain the signal spectrum. Then, use the correlation between Fourier spectrum and power spectrum to calculate the signal power spectrum.

3. **Calculate Mel-scale filter banks:** Due to the stronger recognition ability of the human ear for lower frequency sounds, the power spectrum is passed through a set of Mel-scale triangular filters to extract signal energy from different frequency bands. Among them, the Mel-scale is calculated from the frequency using the following Equation 4.54:

$$f_{mel} = 2595 * lg\left(1 + \frac{f}{700}\right) \qquad (4.54)$$

4. **Calculate MFCC:** Due to the high correlation of the filter bank coefficients obtained through Step 3, discrete cosine transform (DCT) is used to decorrelate the filter bank coefficients and obtain the MFCC. Finally, the transformed low-frequency interval data is extracted as effective MFCC feature parameters.

The chromaticity spectrum is obtained by projecting the entire spectrum of an audio signal onto 12 intervals (sound levels) and accumulating the energy of the same sound level at different octaves. The spectral centroid represents the center of gravity of the frequency components of an audio signal, commonly used to describe the timbre properties of audio, obtained by energy-weighted averaging of signals within a certain frequency range. The spectral bandwidth describes the frequency range that constitutes the audio signal. Spectrum attenuation describes the cutoff frequency of an audio signal, which is obtained through time-frequency analysis of the signal. The zero crossing rate describes the number of times an audio signal passes through zero, including the number of times the signal goes from positive to negative and from negative to positive. It is commonly used in the classification of percussion sounds.

Taking the audio signal with a load of $5N$ as an example, the above characteristics of the signal were calculated, as shown in Figure 4.56.

4.2.3.3 Feature extraction of wear morphology

Due to the large number of sampling points and data volume of surface morphology captured using a white light interferometer, directly using it as the output of the model can lead to low computational efficiency. Therefore, this section utilizes statistical methods to extract the features of surface morphology, providing a data basis for the correlation analysis between multi-source information of friction processes and worn surface morphology.

4.2.3.3.1 Rough feature extraction of wear morphology

The existing morphology characterization methods are mainly divided into roughness characterization and fractal characterization. Among them, the roughness characterization method uses roughness parameters to describe the surface morphology, which can be further divided into characterization for two-dimensional morphology contours and characterization for three-dimensional surface morphology. Due to the inability to reflect the overall state of surface morphology through the characterization of two-dimensional surface profiles, there are certain limitations in describing morphology. Currently, researchers at home and abroad mostly use three-dimensional surface morphology for the characterization and analysis of worn surface morphology. The commonly used characterization parameters include four height parameters, four spatial parameters, three mixed parameters, and six functional parameters, all of which are obtained through statistical analysis of worn surface morphology and roughness. Among them, the skewness in the height parameter S_{sk} and kurtosis S_{ku},

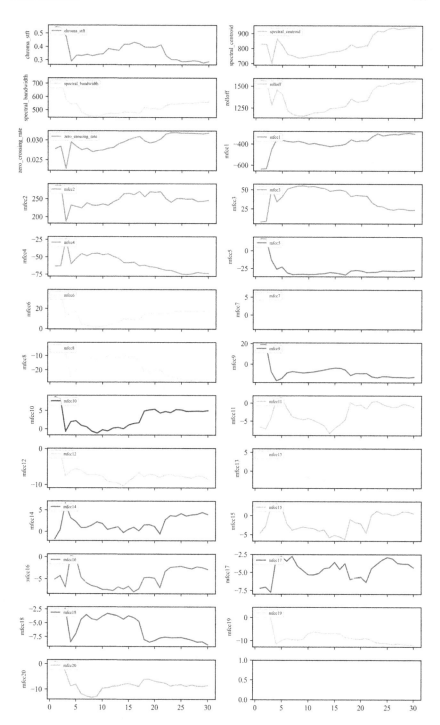

FIGURE 4.56 Audio signal features.

respectively, reflects the peak valley characteristics and sharpness of surface morphology, which has attracted significant attention from researchers. In addition, researchers often combine the root mean square deviation S_q, skewness S_{sk}, and kurtosis S_{ku} of surface morphology height as roughness inputs for morphology simulation when applying rough surface simulation methods to the field of tribology, indicating that these three roughness parameters can effectively describe the characteristics of worn surface morphology. The equations for calculating root mean square deviation, skewness, and kurtosis are as follows:

$$S_q = \sqrt{\frac{1}{MN} \sum_{i=1}^{M} \sum_{j=1}^{N} (z_{i,j} - \bar{z})^2} \qquad (4.55)$$

$$S_{sk} = \frac{\frac{1}{MN} \sum_{i=1}^{M} \sum_{j=1}^{N} (z_{i,j} - \bar{z})^3}{S_q^3} \qquad (4.56)$$

$$S_{ku} = \frac{\frac{1}{MN} \sum_{i=1}^{M} \sum_{j=1}^{N} (z_{i,j} - \bar{z})^4}{S_q^4} \qquad (4.57)$$

Among them, M and N represent the number of sampling points in the two directions of the morphology, $z_{i,j}$ represents the height of the morphology at (i, j), and $\bar{z}$ represents the average height of the morphology.

Taking the worn surface morphology under $5N$ load conditions as an example, Figure 4.57 shows the changes in the morphology and height distribution of the disc surface marked as "2" during the experimental process. From Figure 4.57, it can be seen that before the start of the friction and wear test, the surface of the disc is relatively smooth and the height distribution is relatively concentrated. As the friction test progresses, the surface undulation of the disc gradually becomes apparent, and the height distribution gradually disperses. Meanwhile, by comparing with the Gaussian distribution, the height distribution on the visible disk surface does not meet the Gaussian distribution.

Extract roughness features from the worn surface morphology under $5N$ load conditions and the results are shown in Figure 4.58. The figure shows the changes in the average root mean square deviation, skewness, and kurtosis characteristics of the worn area of the sample disk (averaging the four areas marked on the disk surface) during the test process. It can be seen that as the experiment progresses, the root mean square deviation gradually increases but the growth rate gradually slows down, which is consistent with the changes in the worn surface morphology image in Figure 4.57. The skewness shows a fluctuating decrease and gradually approaches zero, while the kurtosis suddenly increases at the beginning of the friction test and then gradually decreases. This early change is obvious, and the later trend towards stability is

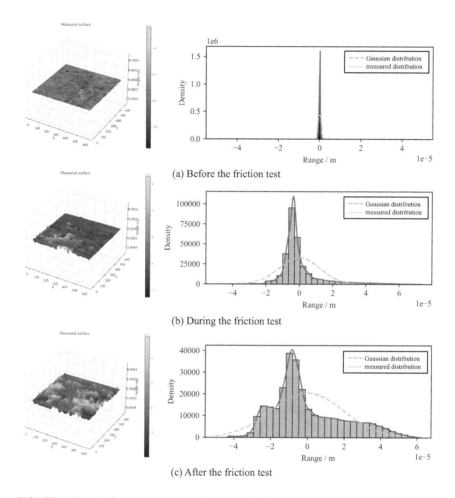

(a) Before the friction test

(b) During the friction test

(c) After the friction test

FIGURE 4.57   Surface topography and height distribution of the sample disk.

consistent with the evolution process of the wear state, indicating that the friction pair has gone through the stage of running in wear and entered the stage of stable wear.

4.2.3.3.2   Periodic feature extraction of wear morphology

In addition to rough features, in order to comprehensively characterize the morphology of the worn surface, it is necessary to introduce spatial characterization parameters, which describe the periodicity of the worn surface height along the X and Y directions. This can be achieved using the two-dimensional autocorrelation function (ACF). For a two-dimensional surface, the two-dimensional ACF is the mathematical expectation of the product of the height of a point on the surface and the height of a point (x, y) away from it. It is described in mathematical language as follows:

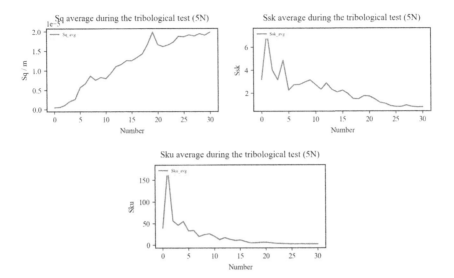

FIGURE 4.58 Evolution of roughness features of worn surface topography during the tribo-test.

$$R(x, \ y) = E\left[z(m, \ n) \times z(m + x, \ n + y)\right] \tag{4.58}$$

In the simulation research of rough surfaces, exponential ACFs are commonly used to fit the two-dimensional ACF of rough surfaces, and the equation is as follows:

$$ACF(x, \ y) = \sigma^2 e^{-2.3\sqrt{\left(\frac{x}{\beta_x}\right)^2 + \left(\frac{y}{\beta_y}\right)^2}} \tag{4.59}$$

Among them, β_x, β_y are the autocorrelation lengths in the x and y directions, respectively, the lengths corresponding to the decay of the autocorrelation values in the x and y directions to 10% of the maximum value; σ is the root mean square deviation of the morphology height.

Therefore, the periodic characteristics of worn surface morphology can be used β_x and β_y to describe. To extract the autocorrelation length value of the worn surface morphology during the experimental process, this chapter calculates the two-dimensional autocorrelation of the surface morphology and uses an exponential ACF to fit it. The two-dimensional autocorrelation calculation is obtained by multiplying the Fourier spectrum conjugation of the surface height based on the similarity between convolution and correlation equations.

Taking the worn surface morphology under $5N$ load conditions as an example, the extracted autocorrelation length variation results are shown in Figure 4.59. Among them, the x-direction corresponds to the circumferential direction of the

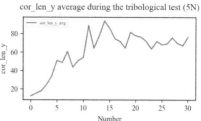

FIGURE 4.59 Evolution of periodic features of worn surface topography during the tribo-test.

sample disk, and the y-direction corresponds to the radial direction of the sample disk. It can be seen that the autocorrelation lengths in both directions show a rapid increase and then a continuous fluctuation trend, indicating that the periodicity of surface morphology in the later stage is relatively stable.

4.2.3.4 Summary

This chapter uses various data processing methods to extract features from the friction process information and wear surface morphology collected during friction and wear tests. First, signal filtering and signal clipping methods were used to preprocess multi-source information of the friction process. Then, universal time-frequency domain features were extracted for all friction process information, and unique signal features were extracted for different friction process information. In addition, roughness features and periodic features were extracted from the worn surface morphology during the friction test process to describe the trend of wear surface morphology changes during the friction test process. The data processing in this chapter lays the foundation for the correlation and fusion of tribological information.

4.2.4 RECONSTRUCTION OF WEAR SURFACE MORPHOLOGY BASED ON MULTI-SOURCE FRICTION INFORMATION FUSION

4.2.4.1 Introduction

Real-time in-situ monitoring of worn surface morphology helps researchers obtain and analyze the friction and wear status of materials, which is a challenge in tribological experimental research. Many existing tribological test systems do not have this function. The feature extraction and preliminary analysis of multi-source information and wear surface morphology in the friction process in previous section have shown that they have a certain correlation and can reflect the wear state of the material to a certain extent. This chapter first introduces the multi-source information fusion method and then uses correlation analysis to determine the correlation between the multi-source information features of the friction process and the wear surface morphology features. Based on this, different machine learning regression models are used to describe the correlation between the two, achieving wear surface morphology feature monitoring based on multi-source information of the friction process. Then, rough surface simulation method is used to reconstruct the wear surface morphology and provide a technical path for real-time in-situ monitoring of worn surface morphology.

4.2.4.2 Morphological feature monitoring based on multi-source information fusion of friction process

Due to the large amount of data collected from multiple sources of friction process information and wear surface morphology, directly using it as model input and output can lead to low computational efficiency. Therefore, this section takes the multi-source friction information features and wear surface morphology features extracted from the previous section as the research object, analyzes the correlation between the two, and constructs the correlation between the two. By integrating easily obtainable and observable friction process information such as force, vibration, sound, and sound pressure, the monitoring of wear surface morphology features is achieved.

4.2.4.2.1 Multi-source information fusion method

The overall framework of the multi-source information fusion method used in this chapter is shown in Figure 4.60. First, the multi-source information of the friction process collected by different sensors is preprocessed and feature extraction. On this basis, the correlation analysis method is used to evaluate the degree of correlation between different friction process information

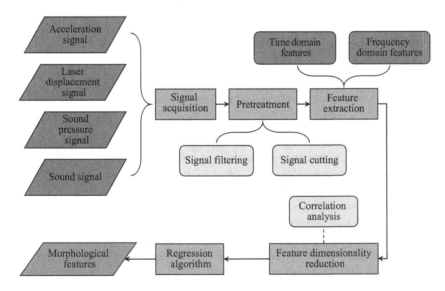

FIGURE 4.60 Multi-source information fusion method.

characteristics and wear surface topography characteristics, providing a theoretical basis for feature dimensionality reduction. Then, the friction process information features from different sensors are normalized or standardized to eliminate the influence of different friction process information, units, and scales of different features. After feature preprocessing, different friction process information within the same time range is fused at the feature level as the input of the wear surface topography reconstruction method.

Compared with the reconstruction of wear surface morphology based on a single friction process information, the wear surface morphology reconstruction method based on multi-source information fusion ensures the integrity of tribological information and can improve the accuracy of wear surface morphology reconstruction.

4.2.4.2.2 Correlation analysis

Data correlation analysis is used to measure the degree of correlation between multiple variables. In the fields of machine learning and deep learning, correlation analysis can provide support for the construction of correlation models. This section uses a correlation coefficient matrix to describe the correlation between different friction process information, as well as between friction process information and wear surface morphology features.

Each element in the correlation coefficient matrix represents the Pearson correlation coefficient between the row variable and the column variable, which represents the degree of correlation between the two variables and can be calculated using the following equation:

$$r(X, \quad Y) = \frac{\sum_{i=1}^{n}(X_i - \bar{X})(Y_i - \bar{Y})}{\sqrt{\sum_{i=1}^{n}(X_i - \bar{X})^2}\sqrt{\sum_{i=1}^{n}(Y_i - \bar{Y})^2}} \tag{4.60}$$

The Pearson correlation coefficient is between $[-1,1]$, and the closer the Pearson correlation coefficient approaches 1, the stronger the linear positive correlation between the two variables. The closer the Pearson correlation coefficient approaches -1, the stronger the linear negative correlation between the two variables.

Taking the average root mean square deviation of the worn surface height under $5N$ load conditions as an example, by calculating the Pearson correlation coefficients between multi-source information features and between multi-source information features and root mean square deviation, 20 multi-source information features with the highest correlation with root mean square deviation were selected. The correlation matrix is shown in Figure 4.61. From the figure, it can be seen that the information with high correlation with root mean square deviation is mainly vibration related information, and the absolute values of the correlation coefficients are all above 0.85, indicating a high correlation between the selected multi-source information features and the root mean square

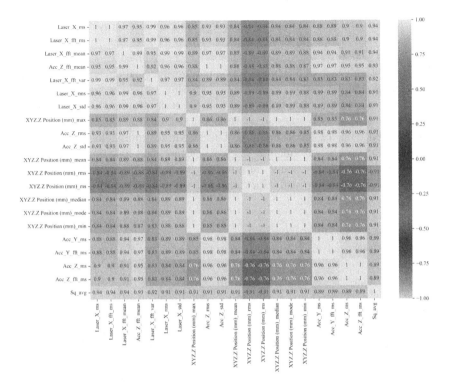

FIGURE 4.61   Correlation coefficient matrix between selected features.

deviation of wear surface height, which can be used for monitoring the root mean square deviation of wear surface height.

4.2.4.2.3 Algorithm for monitoring wear surface morphology features

After selecting the friction process information features that have the highest correlation with the morphology features of each worn surface, a machine learning method is used to construct an association model between the two. The machine learning algorithms used in this chapter include lasso regression, ridge regression, kernel ridge regression (KRR), extreme gradient boosting (XGBoost), SVR, and RF regression.

4.2.4.2.3.1 Lasso regression and ridge regression   Lasso regression and ridge regression, respectively, introduce L1 and L2 regularization terms on the basis of linear regression methods to reduce the overfitting degree of the model. When solving linear regression problems, the least squares method is usually used as a tool. During the solving process, the following objective functions need to be constructed:

$$L(y, \ f(x, \ \omega)) = \sum_{i=1}^{m} (y_i - \omega x_i)^2 \tag{4.61}$$

Among them, ω is the coefficient matrix of the required solution.

On this basis, Lasso regression and ridge regression, respectively, changed the objective function to the following form:

$$L(y, \ f(x, \ \omega)) = \sum_{i=1}^{m} (y_i - \omega x_i)^2 + \alpha \omega_1 \tag{4.62}$$

$$L(y, \ f(x, \ \omega)) = \sum_{i=1}^{m} (y_i - \omega x_i)^2 + \alpha \omega_2^2 \tag{4.63}$$

When training Lasso regression and ridge regression models, the main approach is to adjust α size of the value controls the degree of overfitting of the model. The difference is that Lasso regression can be used for feature selection, which sets the coefficients of features with low model participation to 0.

Due to the large number of multi-source information features extracted in this chapter, it is easy to cause overfitting of the model, which means that the model may perform well in known training set data and poorly in unknown test set data. Therefore, using Lasso regression and ridge regression models can effectively avoid this situation and improve the robustness of wear surface morphology feature monitoring.

4.2.4.2.3.2 Kernel ridge regression   The KRR method is also proposed on the basis of the linear regression method, first introducing a kernel function to transform the objective function into the following form:

$$L(y, \ f(x, \ \omega)) = \sum_{i=1}^{m} (y_i - \psi(x_i))^2 = \sum_{i=1}^{m} (y_i - \sum_{j=1}^{n} \alpha_j \kappa(x_j, \ x_i))^2 \tag{4.64}$$

$\kappa(x_j, x_i)$ represents the kernel function between two training samples, which aims to minimize the error between the weighted sum of the objective value and the kernel function.

In order to avoid overfitting of training data, the KRR method further integrates the ridge regression method and introduces the L1 regularization term, making the objective function in the following form:

$$L(y, \ f(x, \ \omega)) = \sum_{i=1}^{m} (y_i - \sum_{j=1}^{n} \alpha_j \kappa(x_j, \ x_i))^2 + \lambda \sum_{i=1}^{m} \sum_{j=1}^{n} \alpha_i \alpha_j \kappa(x_j, \ x_i) \tag{4.65}$$

Among them, λ is used to balance the accuracy and complexity of the model.

Compared with linear regression methods, KRR method can more effectively handle nonlinear problems and can be used to deal with nonlinear friction and wear evolution problems.

4.2.4.2.3.3 Extreme gradient enhancement decision tree   The XGBoost algorithm, as the name suggests, is a type of boosting algorithm that achieves effective performance improvement by integrating multiple weaker regression tree models together. During the model construction process, the XGBoost algorithm continuously adds trees to fit the residual of the previous prediction. The objective function consists of two parts: One is the error between the predicted value and the true value, and the other is the regularization term to prevent overfitting of the model. The specific equation is as follows:

$$L = \sum_{i=1}^{m} l(y_i, \hat{y}_i) + \sum_{k=1}^{K} (\gamma T + \frac{1}{2}\lambda\omega^2) \qquad (4.66)$$

Among them, T represents the number of child nodes, ω represents the weights of child nodes, using γ and λ to control the number and weight of sub nodes and avoid overfitting of the model.

In the process of building the model, as the newly added tree is mainly used to fit the residual of the previous prediction, the prediction result after adding the t-th tree can be expressed as:

$$\hat{y}_i^{(t)} = \hat{y}_i^{(t-1)} + f_t(x_i) \qquad (4.67)$$

Like the Lasso regression, ridge regression, and KRR algorithms mentioned above, XGBoost introduces regularization terms to better control the overfitting degree of the model. At the same time, XGBoost algorithm has high accuracy and speed in processing medium and low dimensional data in friction and wear research.

4.2.4.2.3.4 Support vector regression   SVR is proposed on the basis of SVM. The main idea of SVM is to map linearly indivisible features to higher dimensional feature spaces, enabling them to be separated by hyperplanes in high-dimensional spaces. Furthermore, linear algorithms are used to analyze the data. The main idea of SVR is to map feature samples to a higher dimensional feature space, find a hyperplane, and minimize the distance between the samples and the hyperplane. The objective function of SVR can be written in the following form:

$$L = \frac{1}{2}\omega^2 + C \sum_{i=1}^{m} l_\varepsilon(f(x_i), \ y_i) \qquad (4.68)$$

Among them, C is the regularization coefficient, ε is the deviation between the acceptable predicted value $f(x)$ and the true value y, therefore l_ε only calculates deviations greater than ε, the part of time.

SVR can effectively deal with high-dimensional input problems, maintaining good performance even when the feature size of the dataset is greater than the sample size. Due to the large amount of multi-source information features extracted and filtered in this chapter, SVR can be used to achieve the desired goals.

4.2.4.2.3.5 Random forest regression   RF regression is a branch of the RF algorithm, proposed on the basis of decision trees, and belongs to a type of bagging algorithm. Similar to the XGBoost model, RF regression, as the name suggests, also consists of multiple weakly performing decision trees. During the construction process, multiple unrelated decision trees are constructed by randomly extracting sample features from the dataset. During the prediction process, the final prediction result is obtained by averaging or voting on the prediction results of all trees. Unlike XGBoost, the different decision trees of the RF regression algorithm remain independent, and there is no inheritance relationship. The RF regression algorithm has good noise resistance and can effectively eliminate the influence of noise in friction and wear test data.

To achieve feature monitoring of worn surface morphology, this chapter utilizes all the features extracted in Chapter 3 to construct a dataset under corresponding operating conditions, which serves as the input and output of the model. First, 20 multi-source information features with the highest correlation were selected based on different wear surface morphology feature parameters, and then 20 features were used as inputs to predict wear surface morphology features using the 6 machine learning models mentioned above. Due to the fact that the number of data samples under each operating condition is 30, during the prediction process, 22 samples are randomly selected as the training set, and the remaining 8 samples are used as the testing set, resulting in a proportion of 25–30% of the total sample size in the testing set. During the construction process of the model, the training set is used for parameter selection and optimization of the model. After completing the construction and parameter optimization of different machine learning models, the model is applied to the test set to obtain the predicted results of the model.

Figures 4.62 and 4.63, respectively, show the optimal prediction results for the roughness characteristic parameters and periodic characteristic parameters of the worn surface morphology in the experimental dataset with a load of 5N. Among them, Figure 4.62(a) shows the optimal prediction results of the root mean square deviation of the worn surface height, which is predicted by the KRR model. The horizontal axis of the left figure is the measured root mean square deviation of the worn surface height, and the vertical axis is the root mean square deviation of the worn surface height predicted by the KRR model. By comparing with the red line $y = x$, the prediction results can be more intuitively analyzed. It can be seen that the predicted points are concentrated near the red straight line, indicating that the predicted and measured results are relatively close. The right figure takes the number of friction and wear test groups as the horizontal axis, while the blue broken line and orange dashed line, respectively, represent the measured and predicted values of the root mean square deviation of the wear surface height. The trend of the two is similar, indicating that the KRR method can excavate the features related to the root mean square deviation of the wear surface height in the multi-source information of the friction process. In order to grasp the trend of the root mean square deviation of the worn surface height gradually increasing during the friction test process. Similarly, Figure 4.62(b)

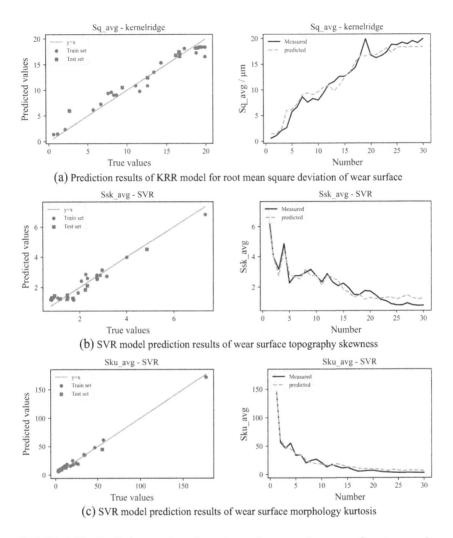

(a) Prediction results of KRR model for root mean square deviation of wear surface

(b) SVR model prediction results of wear surface topography skewness

(c) SVR model prediction results of wear surface morphology kurtosis

FIGURE 4.62 Prediction results of roughness features of worn surface topography under 5*N*.

and (c), respectively, shows the optimal prediction results for the skewness and kurtosis of worn surface morphology, which were both predicted by the SVR algorithm. The results indicate that the SVR algorithm can effectively grasp the trend of changes in skewness and kurtosis of worn surface morphology and has good predictive ability.

Figure 4.63(a) shows the optimal prediction results of the *X*-direction autocorrelation length of the worn surface morphology, which were predicted by the RF regression model. The horizontal axis of the left figure represents the actual measured values, and the vertical axis represents the predicted values obtained using the RF regression model. It can be seen that the prediction point is

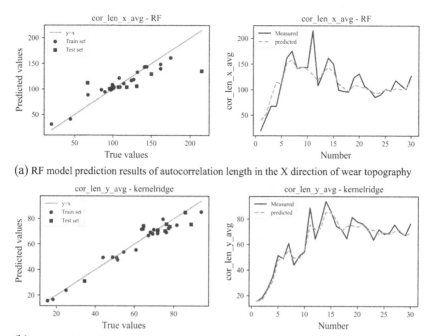

(a) RF model prediction results of autocorrelation length in the X direction of wear topography

(b) KRR model prediction results of autocorrelation length in the Y direction of wear topography

FIGURE 4.63 Prediction results of periodic features of worn surface topography under 5N.

close to the red line $y = x$, and the right figure uses the number of friction and wear test groups as the horizontal axis. The blue broken line and orange dashed line represent the measured and predicted values of the autocorrelation length in the X-direction of the worn surface morphology, respectively. The overall trend of the two is consistent, indicating that the RF regression algorithm can grasp the change process of the autocorrelation length in the X-direction of the worn surface morphology during friction testing from a global perspective. Similarly, Figure 4.63(b) shows the optimal prediction results of the Y-direction auto-correlation length of the worn surface morphology, which were predicted by the KRR algorithm. Compared with the prediction results of the X-direction autocorrelation length, the KRR algorithm can effectively track the Y-direction autocorrelation length of the worn surface morphology during friction testing, and the prediction accuracy is higher.

In order to quantitatively analyze the predictive performance of the model, it is necessary to evaluate the model. This chapter selects indicators such as the coefficient of determination (R2), mean square error (MSE), and root mean square error (RMSE) to evaluate the prediction results of different machine learning models for different wear surface morphology features. The calculation equations are as follows:

$$R^2 = 1 - \frac{\sum_{i=0}^{m} (y_i - \hat{y}_i)^2}{\sum_{i=0}^{m} (y_i - \bar{y})^2} \tag{4.69}$$

$$MSE = \frac{1}{m} \sum_{i=0}^{m} (y_i - \hat{y}_i)^2 \tag{4.70}$$

$$RMSE = \sqrt{MSE} = \sqrt{\frac{1}{m} \sum_{i=0}^{m} (y_i - \hat{y}_i)^2} \tag{4.71}$$

The closer the determination coefficient R^2 value is to 1, the better the prediction performance of the model. The smaller the mean square error and root mean square error values, the smaller the prediction error of the model.

Based on the datasets obtained from the basic friction and wear tests conducted in this chapter under three different load conditions, the above three evaluation indicators were calculated for the prediction results of wear surface morphology characteristics of the test set. The results are shown in Tables 4.15, 4.16, and 4.17. Overall, KRR, SVR, and RF regression models can effectively establish the

TABLE 4.15
Prediction Results of Worn Surface Topography Features Under 5N

Wear Surface Topography Characteristics	Model	R^2	MSE	RMSE
RMSD S_q	**KRR**	**0.930**	**1.773 μm^2**	**1.331 μm**
	XGBoost	0.800	5.042 μm^2	2.245 μm
	SVR	0.915	2.149 μm^2	1.466 μm
	RF	0.924	1.917 μm^2	1.385 μm
	Lasso	0.905	2.410 μm^2	1.553 μm
	Ridge	0.890	2.789 μm^2	1.670 μm
Skewness S_{sk}	KRR	0.601	0.543	0.737
	XGBoost	0.556	0.603	0.777
	SVR	**0.921**	**0.108**	**0.328**
	RF	0.595	0.550	0.742
	Lasso	0.483	0.703	0.838
	Ridge	0.517	0.656	0.810
Kurtosis S_{ku}	KRR	0.709	76.643	8.755
	XGBoost	0.424	151.566	12.311
	SVR	**0.912**	**23.181**	**4.815**
	RF	0.709	76.565	8.750
	Lasso	0.743	67.669	8.226
	Ridge	0.487	134.978	11.618
X-direction autocorrelation length β_x	KRR	0.289	1242.371	35.247
	XGBoost	0.210	1380.994	37.162
	SVR	0.266	1283.573	35.827

TABLE 4.15 (Continued)
Prediction Results of Worn Surface Topography Features Under 5N

Wear Surface Topography Characteristics	Model	R^2	MSE	RMSE
	RF	**0.311**	**1203.649**	**34.694**
	Lasso	0.278	1261.814	35.522
	Ridge	0.306	1213.606	34.837
Y-direction autocorrelation	**KRR**	**0.837**	**40.551**	**6.368**
length β_y	XGBoost	0.730	67.197	8.197
	SVR	0.620	94.776	9.735
	RF	0.631	91.874	9.585
	Lasso	0.758	60.247	7.762
	Ridge	0.730	67.364	8.208

TABLE 4.16
Prediction Results of Worn Surface Topography Features Under 7.5N

Wear Surface Topography Characteristics	Model	R^2	MSE	RMSE
RMSD S_q	KRR	0.767	1.196 μm²	1.094 μm
	XGBoost	0.784	1.108 μm²	1.053 μm
	SVR	0.720	1.436 μm²	1.198 μm
	RF	0.611	1.994 μm²	1.412 μm
	Lasso	**0.800**	**1.025 μm²**	**1.012 μm**
	Ridge	0.667	1.709 μm²	1.307 μm
Skewness S_{sk}	KRR	0.753	0.160	0.400
	XGBoost	0.774	0.146	0.382
	SVR	0.411	0.381	0.617
	RF	0.711	0.187	0.433
	Lasso	0.803	0.127	0.357
	Ridge	**0.805**	**0.126**	**0.355**
Kurtosis S_{ku}	**KRR**	**0.827**	**0.259**	**0.509**
	XGBoost	0.526	0.710	0.843
	SVR	0.789	0.316	0.562
	RF	0.819	0.271	0.521
	Lasso	0.784	0.324	0.569
	Ridge	0.795	0.307	0.554
X-direction autocorrelation	**KRR**	**0.901**	**321.845**	**17.940**
length β_x	XGBoost	0.849	489.328	22.121
	SVR	0.830	550.473	23.462
	RF	0.607	1273.579	35.687

(Continued)

TABLE 4.16 (Continued)
Prediction Results of Worn Surface Topography Features Under 7.5N

Wear Surface Topography Characteristics	Model	R^2	MSE	RMSE
	Lasso	0.837	528.672	22.993
	Ridge	0.880	387.905	19.695
Y-direction autocorrelation	KRR	0.516	138.439	11.766
length β_y	**XGBoost**	**0.798**	**57.818**	**7.604**
	SVR	0.098	258.013	16.063
	RF	0.732	76.642	8.755
	Lasso	0.176	235.734	15.354
	Ridge	0.113	253.855	15.933

TABLE 4.17
Prediction Results of Worn Surface Topography Features Under 10N

Wear Surface Topography Characteristics	Model	R^2	MSE	RMSE
RMSD S_q	KRR	0.964	0.849 μm^2	0.921 μm
	XGBoost	**0.976**	**0.561 μm^2**	**0.749 μm**
	SVR	0.897	2.447 μm^2	1.564 μm
	RF	0.952	1.134 μm^2	1.065 μm
	Lasso	0.936	0.961 μm^2	0.980 μm
	Ridge	0.960	0.961 μm^2	0.980 μm
Skewness S_{sk}	KRR	0.772	0.221	0.470
	XGBoost	0.799	0.194	0.441
	SVR	0.487	0.497	0.705
	RF	**0.828**	**0.166**	**0.408**
	Lasso	0.489	0.495	0.704
	Ridge	0.470	0.513	0.716
Kurtosis S_{ku}	KRR	0.856	3.959	1.990
	XGBoost	0.440	15.394	3.924
	SVR	0.604	10.882	3.299
	RF	**0.948**	**1.431**	**1.196**
	Lasso	0.401	16.478	4.059
	Ridge	0.766	6.435	2.537
X-direction autocorrelation	KRR	0.790	387.578	19.687
length β_x	XGBoost	0.595	746.746	27.327
	SVR	**0.815**	**340.515**	**18.453**
	RF	0.482	955.010	30.903
	Lasso	0.709	535.800	23.147
	Ridge	0.752	458.279	21.407

TABLE 4.17 (Continued)
Prediction Results of Worn Surface Topography Features Under 10N

Wear Surface Topography Characteristics	Model	R^2	MSE	RMSE
Y-direction autocorrelation length β_y	**KRR**	**0.763**	**69.311**	**8.325**
	XGBoost	0.707	85.871	9.277
	SVR	0.664	98.460	9.923
	RF	0.697	88.772	9.422
	Lasso	0.603	116.157	10.778
	Ridge	0.683	92.924	9.640

correlation between multi-source information of friction processes and wear surface morphology features, and the prediction effect of different wear surface morphology feature parameters is generally good. As shown in Table 4.15, under the test condition with a load of 5N, except for the autocorrelation length in the X-direction of the worn surface morphology, the R^2 values of all other feature parameters are close to 1, and the MSE and RMSE are relatively small, indicating excellent prediction performance of the model. Among them, the KRR model and SVR model achieved the best performance in predicting the two feature parameters of the worn surface morphology, respectively. This indicates that these two models have a wide range of applicability. Regarding the prediction of the autocorrelation length in the X-direction of the worn surface morphology, it can be seen from Figure 4.63(a) that although the specific predicted values have a significant deviation, the RF regression method can better grasp the trend of feature changes in the friction and wear process, providing support for monitoring the wear status. As shown in Table 4.16, under the test condition with a load of 7.5N, the predicted results of the five characteristic parameters of worn surface morphology are relatively balanced, with R^2 values ranging from 0.8 to 0.9, and MSE and RMSE are relatively small. Among them, the KRR method has achieved the best performance in predicting the two characteristic parameters of worn surface morphology. As shown in Table 4.17, under the experimental condition of a load of 10N, the prediction results of the five characteristic parameters of the worn surface morphology are also relatively balanced. Except for the optimal prediction R^2 value of the Y-direction autocorrelation length of the worn surface morphology, which is 0.763, the optimal R^2 values of the other parameters reach 0.8–1.0. Among them, the RF regression model achieved the optimal prediction of the two characteristic parameters of the worn surface morphology.

In summary, for the prediction of wear surface morphology feature parameters under different load conditions, using different machine learning models can achieve high accuracy, indicating that these models can effectively mine and extract wear-related information hidden in multi-source signals during the friction process, providing a foundation for real-time in-situ monitoring of wear surface morphology.

4.2.4.3   Reconstruction of worn surface morphology

The morphology of the worn surface intuitively reflects the characteristics and state of the surface during the friction and wear process of materials. Monitoring and analyzing the morphology of the worn surface is helpful for researchers to explore and reveal the mechanism and essence of contact and friction wear between materials. In order to achieve real-time reconstruction of wear surface morphology, various machine learning models have been used in previous sections to construct the correlation between wear surface morphology features and multi-source information of friction processes. Based on this, it is necessary to construct a model with wear surface morphology features as input and 3D wear morphology as output. Therefore, this section applies the simulation method of random rough surface morphology to the reconstruction of worn surface morphology, providing a complete technical path for real-time in-situ monitoring of worn surface morphology in friction and wear experimental research.

4.2.4.3.1   Method for reconstructing worn surface morphology

At present, there have been relevant studies both domestically and internationally applying rough surface simulation methods to the field of tribology as input for simulation research. The simulation of rough surfaces can be achieved through the construction of time series models, among which the moving average (MA) model is widely used. From the perspective of statistical distribution, rough surfaces can be divided into two types: Gaussian rough surfaces and non-Gaussian rough surfaces. From the surface height distribution of the sample disk in the friction and wear test in Figure 4.57, it can be seen that the worn surface of the sample disk belongs to non-Gaussian rough surfaces. The most commonly used method in foreign countries currently is to use the Johnson transformation system to transform Gaussian random sequences into non-Gaussian sequences that meet the statistical parameters requirements of target skewness and kurtosis, in order to construct a three-dimensional surface with expected skewness and kurtosis.

This chapter takes the predicted wear surface morphology features as input and reconstructs the wear surface morphology according to the process shown in Figure 4.64, providing support for real-time reconstruction of wear surface morphology.

The overall idea of the method for reconstructing worn surface morphology is as follows:

First, use the MA model to model the target surface:

$$z_{i,j} = \sum_{k=-\frac{N}{2}}^{\frac{N}{2}} \sum_{l=-\frac{M}{2}}^{\frac{M}{2}} a_{k,l}\, \eta_{i+k,\,j+l} \tag{4.72}$$

Where z is the height of the target surface, η is a random non-Gaussian sequence, and a is the corresponding coefficient matrix.

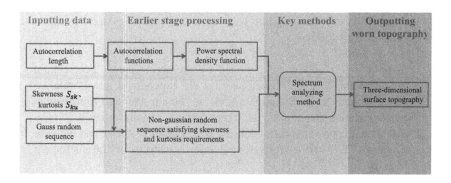

FIGURE 4.64 Reconstruction method of worn surface topography.

The generation of random non-Gaussian sequences can be achieved based on the Johnson transformation system. By inputting the target skewness and kurtosis, the skewness and kurtosis of the input non-Gaussian sequence can be calculated. By applying the calculated skewness and kurtosis to the random Gaussian sequence using the Johnson transformation system, a random non-Gaussian sequence that satisfies the target distribution can be obtained as the input sequence of the MA model.

For the calculation of coefficient matrix a, first clarify the following relationship between the two-dimensional autocorrelation function and the coefficient matrix:

$$ACF(p, \quad q) = \sum_{k=-\frac{N}{2}+p}^{\frac{N}{2}} \sum_{l=-\frac{M}{2}+q}^{\frac{M}{2}} a_{k, \ l} a_{k-p, \ l-q} \tag{4.73}$$

By utilizing the characteristic that the power spectral density (PSD) function is a Fourier transform of the autocorrelation function, Fourier transforms are performed on the left and right sides of the above equation to obtain:

$$PSD(\omega_p, \quad \omega_q) = H(\omega_p, \quad \omega_q) \times \bar{H}(\omega_p, \quad \omega_q) \tag{4.74}$$

Among them, H is the Fourier transform of a. Due to the fact that the power spectral density is composed of real coefficients, H can be obtained by transforming from the above equation:

$$H(\omega_p, \quad \omega_q) = \sqrt{PSD(\omega_p, \quad \omega_q)} \tag{4.75}$$

Perform Fourier transform on the left and right sides of Equation (4.72), and substitute Equation (4.75) to obtain the Fourier transform of rough surface height. By performing inverse Fourier transform on it, the target rough surface can be obtained.

The steps for reconstructing the worn surface morphology used in this case are summarized as follows:

1. Substitute β_x, β_y that are obtained by predicting the autocorrelation length in the X and Y directions into the exponential autocorrelation function shown in Equation (4.58) to obtain the two-dimensional autocorrelation function of the target surface, and discretize it.
2. Calculate the power spectral density using Equation (4.74), and then use Equation (4.75) to calculate the Fourier transform of the coefficient matrix.
3. Input a random Gaussian sequence and the skewness and kurtosis of the target surface morphology, and obtain a random non-Gaussian sequence with specified skewness and kurtosis based on the Johnson transformation system.
4. Perform Fourier transform on the generated random non-Gaussian sequence, and multiply it with the Fourier transform of the coefficient matrix. Perform inverse Fourier transform on the obtained results to obtain the final three-dimensional surface morphology height matrix.

4.2.4.3.2 *Verification of reconstruction effect*

To verify the effectiveness of the method for reconstructing worn surface morphology, this chapter extracted the surface morphology of four regions of the sample disk at the same time, as shown in Figure 4.65. Using the average values of morphology features in four regions as the target parameter, the three-dimensional morphology at that time was reconstructed, as shown in Figure 4.66. From the three-dimensional surface map on the left side of Figure 4.65(b) and (e), it can be seen that the surface morphology of the sample disk at this time is generally relatively flat, with some small protrusions distributed relatively evenly on the surface. From the surface height distribution map on the right side of Figure 4.65(b) and (e), it can be seen that the surface height distribution at this time is sharper and steeper than the Gaussian distribution under the same standard deviation parameter, indicating that the surface morphology at this time has a larger kurtosis. At the same time, there are some slight differences in the surface morphology and height distribution of different areas of the same sample disk. On the left side of Figure 4.66 is a three-dimensional surface map reconstructed using the average values of the morphology features of the four regions shown in Figure 4.6, which can be used to approximately describe the overall morphology of the wear area of the sample disk. Similarly, the simulated three-dimensional morphology is relatively flat, with some slight protrusions distributed on the surface. On the right side of Figure 4.66 is the corresponding surface height distribution map, which is compared to the surface height distribution map on the right side of Figure 4.65. The visible simulated three-dimensional morphology describes the average state of height distribution in four regions.

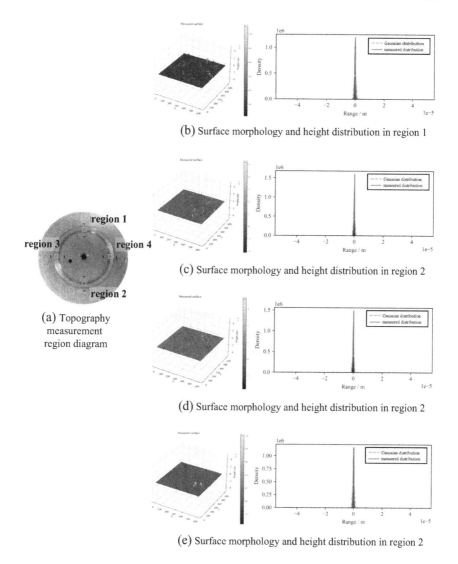

(b) Surface morphology and height distribution in region 1

(c) Surface morphology and height distribution in region 2

(d) Surface morphology and height distribution in region 2

(e) Surface morphology and height distribution in region 2

(a) Topography measurement region diagram

FIGURE 4.65 Surface topography and height distribution of the four regions of the disc.

In order to quantitatively analyze the effectiveness of the wear surface morphology reconstruction method used in this chapter, the characteristic parameters of the actual surface morphology and corresponding simulated surface morphology of four regions were calculated to describe the similarity between the simulated surface morphology and the actual surface morphology. Table 4.18 shows the roughness and periodic characteristic parameters of the calculated measured and simulated morphologies. It can be seen that the simulated surface morphologies describe the average attributes of the actual

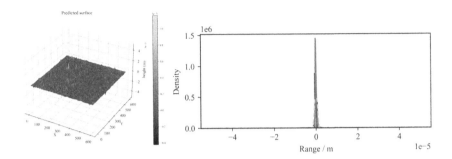

FIGURE 4.66 Simulated surface topography and height distribution of the disc.

TABLE 4.18

Comparison of Features Between Simulated Surface Topography and Measured Surface Topography

Topography Feature Parameters	Measured Topography (Region 1)	Measured Topography (Region 2)	Measured Topography (Region 3)	Measured Topography (Region 4)	Simulated Topography
RMSD $S_q/\mu m$	0.59	0.37	0.35	0.52	0.46
Skewness S_{sk}	4.28	1.51	0.21	3.82	2.53
Kurtosis S_{ku}	44.73	19.52	7.36	58.21	32.04
X-direction autocorrelation length	14	11	11	15	12
Y-direction autocorrelation length	13	12	10	14	11

surface morphologies from the perspectives of roughness and periodic characteristics, which can be used to characterize the overall surface state of the wear area of the specimen disk at that time. This further provides a visual tool for real-time in-situ monitoring of wear surface morphologies during friction and wear testing.

To verify the effectiveness of the wear surface morphology reconstruction method based on multi-source friction information fusion, the actual wear surface morphology features and predicted wear surface morphology features in Section 4.2.3 were used as inputs to construct a three-dimensional surface morphology and compare them. Taking the surface morphology under 5N load conditions as an example, Figures 4.67 and 4.68, respectively, show the comparison between the wear surface morphology constructed using real and predicted feature parameters in the early and late stages of friction testing under 5N load conditions.

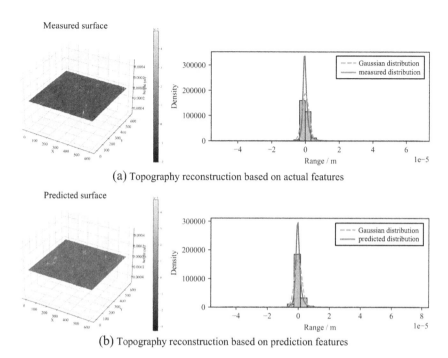

(a) Topography reconstruction based on actual features

(b) Topography reconstruction based on prediction features

FIGURE 4.67   Verification of surface topography reconstruction of the disc in the early stage of the tribo-test.

From the three-dimensional surface morphology shown on the left side of Figure 4.67(a), it can be seen that the surface of the sample disk reconstructed using real feature values in the early stage of the friction test is relatively smooth, with slight overall fluctuations. From the height distribution map of the surface shown on the right side of Figure 4.67(a), it can be seen that the height distribution of the surface is relatively concentrated and balanced. Correspondingly, the surface morphology of the sample disk reconstructed based on predicted eigenvalues shown in Figure 4.67(b) is consistent with Figure 4.67(a) from a visual perspective, and the overall trend of the height distribution map also maintains good similarity.

From the three-dimensional surface morphology shown on the left side of Figure 4.68(a), it can be seen that the surface of the sample disk constructed using real feature values in the later stage of the friction test fluctuates significantly, which is rough compared to the surface morphology of the sample disk in the early stage of the test. From the height distribution diagram of the surface shown on the right side of Figure 4.68(a), it can be seen that the height distribution of the sample disk surface in the later stage of the experiment is relatively dispersed and has a significant degree of deviation. Correspondingly, the surface morphology of the sample disk constructed based on predicted feature values shown in Figure 4.68(b) maintains good consistency with Figure 4.68(a).

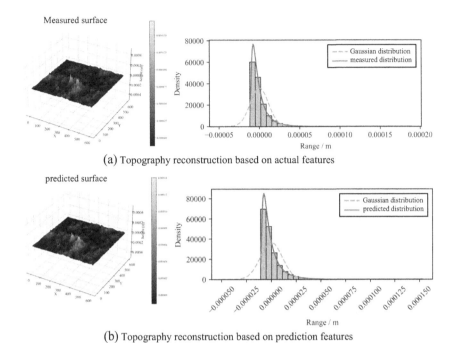

(a) Topography reconstruction based on actual features

(b) Topography reconstruction based on prediction features

FIGURE 4.68 Verification of surface topography reconstruction of the disc in the later stage of the tribo-test.

The above analysis indicates that the wear surface morphology reconstruction method based on multi-source information fusion proposed in this chapter can effectively capture the multi-source information features and key features of wear surface morphology during friction testing, providing a foundation for real-time in-situ monitoring of wear surface morphology. Meanwhile, compared to using only parameters to describe the worn surface morphology, the visualization of worn surface morphology can further introduce details of the morphology, improve its authenticity, and provide model input for constructing a digital twin system for friction and wear experiments. On this basis, the constructed three-dimensional surface morphology can be used as simulation input to avoid long-term testing, save labor and material costs, and provide support for the life assessment and prediction of key components of mechanical equipment.

4.2.4.4 Summary

This chapter first uses correlation analysis method to determine the correlation between multi-source information features of friction process and wear surface morphology features, providing a theoretical basis for feature monitoring of wear surface morphology, and providing a basis for reasonable feature selection. Subsequently, multiple machine learning methods were used to construct the correlation between the screened features and the wear surface morphology features, and the prediction accuracy was compared. The results showed that the KRR method and SVR method had high accuracy. On this basis, a rough surface simulation method was introduced to achieve wear surface morphology reconstruction based on wear morphology features. Thus, a closed-loop real-time reconstruction of worn surface morphology is achieved.

4.2.5 SUMMARY OF THE CASE

Wear and tear problems widely exist in the industrial field and are one of the key factors for mechanical equipment failures or failures. The morphology of worn surfaces can intuitively reflect the characteristics of material wear status, and real-time in-situ monitoring of worn surface morphology can assist researchers in obtaining real-time material wear status, thereby providing support for equipment fault diagnosis and predictive maintenance. First, a multi-source information collection system for friction process was established on a basic friction and wear testing machine, achieving the acquisition of multi-source friction process information. The key features of friction process multi-source information and wear surface morphology were extracted and analyzed, and the reconstruction of wear surface morphology based on multi-source friction information was achieved, providing a foundation for interdisciplinary information association in tribology research. The main research work and conclusions of the case are as follows:

1. Design and data acquisition method of a multi-source information collection system for friction processes

   Applying environment-based design theory to the field of tribology experimental research, a multi-source information collection system for friction process integrating multiple sensors was designed and deployed based on the Rtec-5000S multifunctional friction and wear testing machine. On this basis, friction and wear tests of pin plate foundation under different working conditions were designed and carried out, achieving real-time collection of multi-source information such as vibration, sound, and sound pressure during the friction test process. At the same time, a friction test plan with multiple starts and stops of a single sample was designed to achieve in-situ collection of worn surface morphology, providing a data foundation for subsequent multi-source information fusion and analysis.

2. A method for extracting features of friction process from multiple sources and wear surface morphology

   In response to the multi-source information of the friction process collected during the friction and wear test process, various temporal and frequency domain features were extracted using digital signal processing. At the same time, features commonly concerned in acoustic signal research were extracted for specific sound pressure and sound signals as inputs to the multi-source information fusion model. Based on the rough and periodic characteristics of the worn surface morphology captured by a white light interferometer during the friction and wear test, key features of the worn surface morphology were extracted as the output of a multi-source information fusion model. Through preliminary analysis of the changing trends of multi-source information features and wear surface morphology features in the friction process during the experimental process, it intuitively reflects the correlation between the two and reflects the evolution trend of wear status at the same time.

3. Wear surface morphology monitoring method based on multi-source information fusion of friction process
   Based on the extracted multi-source information features of friction process and wear surface morphology features, correlation analysis was used to determine the linear correlation between different information. On this basis, feature screening was achieved, providing a data basis for monitoring wear surface morphology features. Subsequently, multiple machine learning methods were used to establish an association model between multi-source information of the friction process and the morphology characteristics of the worn surface. The model's prediction effect was verified by comparing experimental data. At the same time, the introduction of rough surface simulation method has achieved feature-based reconstruction of wear surface morphology, achieving real-time closed-loop reconstruction of wear surface morphology through multi-source information fusion in the friction process.
   The main innovation points of this case are as follows:
   1. A multi-source information collection system for friction process based on a universal friction and wear testing machine was designed and validated. Traditional tribological experimental research mainly focuses on tribological information that directly characterizes the wear state of materials, such as friction force, friction coefficient, and wear amount, while neglecting the multi-disciplinary nature of tribology. This chapter applies environmental design theory to tribology research, and designs and builds a multi-sensor integrated multi-source information collection system for friction processes, achieving real-time and in-situ collection of multi-source information such as force, vibration, and acoustics.
   2. A method for reconstructing worn surface morphology was proposed and validated. The real-time in-situ acquisition of sample surface morphology in friction and wear tests faces problems such as small sample size and relative motion between friction pairs. This chapter analyzes the correlation between multi-source information in the friction process and wear surface morphology, and constructs a wear surface morphology reconstruction model based on the fusion of multi-source information in the friction process. This has reference significance for phenomenon analysis and mechanism mining in tribological experimental research.

4.3 REAL-TIME COLLECTION AND TIME SERIES ANALYSIS METHOD FOR MULTI-DIMENSIONAL FRICTION INFORMATION

Friction and wear have a significant impact on the reliability and lifespan of high-end equipment in fields such as rail transit and aerospace. Real-time monitoring of friction and wear has important research value and broad application prospects for equipment operation and maintenance under complex conditions. Real-time monitoring of the friction coefficient of moving components in friction and wear systems is a challenge. The development of intelligent perception and data technology has provided new methods for the acquisition, processing, and application of friction process information. The development of tribology has also entered Tribology 4.0 and developed friction informatics as a new direction of tribology. The friction informatics framework effectively integrates multi-dimensional data such as acoustics, vibration, and images into tribological research such as condition monitoring and prediction of friction and wear behavior.

This case aims to utilize the correlation characteristics between friction information such as sound and vibration during friction and wear tests and friction coefficient and to achieve real-time monitoring of friction coefficient during friction and wear processes. The main research work of this chapter is as follows:

1. We have built a real-time data collection platform for multi-dimensional friction information, achieved real-time data collection of multi-dimensional friction information, and established a data foundation for friction information research. A real-time data collection platform for multi-dimensional friction information such as friction sound, friction vibration, and friction images was built based on a friction and wear testing machine and various sensing devices. Through pretesting, experimental combinations with obvious friction phenomena and significant friction information characteristics were selected, and pin disk friction and wear tests were designed and conducted, obtaining raw data of multi-dimensional friction information.

2. We extracted data features from multi-dimensional friction information, and combined with the variation rules of friction processes, conducted correlation analysis on data distribution and change trends. Based on the original data of multi-dimensional friction information, digital signal processing and feature engineering methods were used to extract the changing features of friction information with tribological research value. In particular, the motion amplification algorithm was used to effectively observe and extract features of the radial micro vibration phenomenon of the upper sample, achieving data fusion of multi-dimensional friction information.

3. An integrated regression model was established to fit the friction coefficient with multi-dimensional friction information, achieving

effective fitting of the friction coefficient, and verifying the stability of the model's fitting effect on multiple working condition test data. A cross-sectional dataset of friction information was formed using the time series cross-sectional method, and multiple basic regression models were selected. An integrated regression model with K-fold cross-validation and double-layer stacking was established, and evaluation indicators for range evaluation were defined. The model was tested using multiple sets of experimental data, achieving accurate fitting of friction coefficients.

This case integrated multi-dimensional friction information to form a time cross-sectional friction information dataset. Through data-driven modeling method, an effective model for fitting multi-dimensional friction information to friction coefficient is established, and the correlation characteristics of friction information are extracted. The accuracy and stability of friction and wear test information data for different working conditions are good, providing a new method for real-time collection and monitoring of friction coefficient.

4.3.1 RESEARCH BACKGROUND OF REAL-TIME ACQUISITION AND TIME SERIES ANALYSIS METHOD OF MULTI-DIMENSIONAL FRICTION INFORMATION

Friction is a common physical phenomenon in production and daily life, which is the interaction between surfaces in relative motion. The friction process can also induce phenomena such as sound, heat, light, and electricity, and the data collected for these phenomena is called friction information. The friction process is the main source of friction information such as friction noise, friction images, and friction vibrations.

The phenomenon of friction and wear and its induced information are widely present in production and daily life, as shown in Figure 4.69. The piercing and screaming noise generated by subway trains, cars, and other transportation vehicles during braking, the photothermal phenomenon during industrial machine tool grinding, and the surface traces left by aircraft during operation are all intuitive manifestations of friction information. These friction information are closely related to the friction phenomena during the operation of various mechanical systems and also brings many problems and challenges to production and daily life.

In the field of rail transit, friction noise is one of the important sources of ecological environmental noise pollution, which seriously affects people's comfort in life and travel. Installing soundproof walls and other facilities next to track facilities can to some extent avoid noise interference with the external environment, but it does not solve the problem of frictional noise from the root. By studying the correlation between friction information such as friction noise and friction wear phenomena, effective control of friction phenomena with negative effects is of great significance for improving the production and living environment.

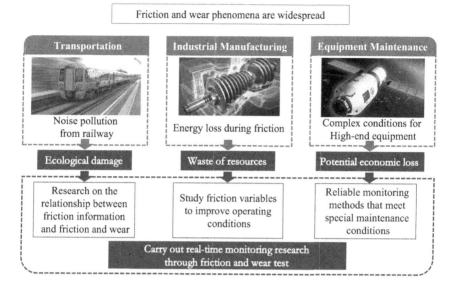

FIGURE 4.69 Research background.

In the field of industrial production, about 30% of the energy generated by the industrial sector is consumed in the friction process, resulting in extremely large consumption of resources such as material and energy. By studying the correlation characteristics between friction information, timely and effective monitoring of friction and wear status can be achieved, thereby improving the operating conditions of industrial equipment and reducing energy and material losses during the friction process. This is also an important goal of green tribology.

In the field of aerospace, gas turbines work in extreme environments and complex conditions such as high temperatures and pressures for a long time, and spacecraft such as satellites need to operate in space for a long time. In high-end equipment represented by this, some friction components are small and critical, and maintenance technology is difficult. Maintenance costs are high, and the potential economic losses of friction failure are huge. Therefore, studying the correlation of friction and wear states to obtain a fast, convenient, and accurate friction state monitoring method that meets extreme conditions is particularly important for improving equipment reliability and reducing significant economic and property losses.

Tribology is a discipline that studies the interaction, changes, and related theories and practices between two relatively moving surfaces in the process of friction and wear, including the study of friction, wear, and lubrication phenomena. The phenomenon of tribology occurs in the surface layer, and its influencing factors are complex and involve multiple disciplines, making theoretical analysis and experimental research more difficult. Therefore, theoretical and experimental research mutually promote and supplement each other, becoming a significant feature of tribological research. As a highly practical technical foundational discipline, the research and testing of tribology are inseparable. Friction and wear testing is an important method for studying the process and state of friction and wear, and friction and wear testing machines are the main instruments for obtaining friction test data.

Creating the same conditions for practical problems to reproduce production problems in friction and wear tests is the most ideal method for studying friction and wear tests. However, in real research, it is often difficult to provide an equivalent experimental environment for friction and wear tests, and the production process is also difficult to meet the precise requirements of data collection under laboratory conditions. This puts forward new requirements for friction and wear test research: On the one hand, by exploring the general laws of friction and wear through basic friction and wear tests, it establishes a foundation for the study of friction and wear phenomena in complex environments and specific conditions. On the other hand, in the friction and wear test, research methods that meet the requirements of the production environment and have strong practicality are adopted, thereby creating opportunities for the promotion of friction and wear research in production and daily life.

In summary, friction phenomena have brought new challenges to the operation and maintenance of equipment under complex environmental conditions. Through tribological research, improving the convenience and accuracy of monitoring the status of friction and wear processes is of great significance for improving production and daily life. The collaborative development of friction and wear experiments and friction theory has laid a solid foundation for the study of the correlation between friction information and friction and wear phenomena.

Tribology research has significant characteristics of combining theory and practice. Relevant research not only helps to improve the operational reliability of industrial equipment and improve the level of equipment monitoring and maintenance technology at the practical level, but also contributes to interdisciplinary research, with significant economic and academic value.

1. Economic significance

Friction failure is one of the important causes of industrial failures, and research has shown that excessive friction and wear are the key failure forms of friction and wear systems. In the industrial field, friction-related failures are a common cause of catastrophic damage to mechanical systems, leading to significant maintenance costs and production delays worldwide. It is estimated that in 2016 alone, the financial loss caused by this in the UK exceeded 500 million euros, equivalent to 0.056% of the gross domestic product. However, industrial applications often lack specific environments and equipment similar to friction and wear experiments, making it difficult to monitor the friction and wear status of industrial equipment in real time. The lag in system data acquisition has brought hidden dangers to production safety and equipment maintenance. Therefore, using basic friction and wear tests to study the friction and wear process, obtain friction information, and achieve system friction state and feature recognition has become a key issue that needs to be urgently solved in friction information research and application

The characteristic of the failure of friction and wear systems is that faults are often sudden. As time goes by, the friction power ages, and the tribological characteristics of the friction and wear system show significant changes. The sudden failure of the friction and wear system is related to the time-varying characteristics of the system. Therefore, real-time collection of friction and wear processes and data analysis from a time series perspective are helpful in studying the changes in the state of friction and wear during the test cycle, thus achieving continuous and effective prevention of friction faults in mechanical systems, and reducing economic losses in industrial production.

2. Academic significance

The study of tribology involves all issues related to energy and material transfer and consumption in mechanical equipment. The friction and wear process generates a large amount of data information, and its potential correlation characteristics have always been a research hotspot in the field of tribology and a key goal of data research in friction and wear test systems.

As shown in Figure 4.70, the friction and wear test system involves four types of data: input variables, environmental variables, output quantities, and

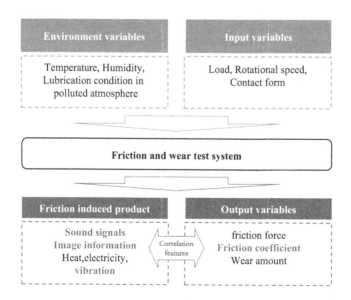

FIGURE 4.70   Friction and wear test system.

friction-induced products. Among them, input variables and environmental variables are the control variables in the experiment; friction-induced products are friction information; output variables are friction, friction coefficient, and wear amount directly related to the state of the friction and wear system; and output variables are the main target data in friction and wear research.

Generally speaking, the input variables are set values, so the amount of data is small and easy to analyze. Traditional friction and wear test research often focuses on the data relationship between the input variables and output variables of the system, forming basic research on the laws of friction and wear states. Multiple research results have become the basic knowledge of friction and wear theory.

It is urgent to conduct on-site and implemented monitoring research on unstable and nonlinear friction and wear processes. Studying the correlation characteristics between friction information and target values such as friction coefficient has important economic and academic value. The development of information technology and the establishment of a friction informatics framework have made it possible to study real-time collected friction information, providing support for studying the state characteristics of friction and wear systems using friction and wear test data.

4.3.2 FRICTION AND WEAR TEST AND REAL-TIME DATA COLLECTION

4.3.2.1 Introduction

Real-time data collection during friction and wear testing is the research foundation for friction process information. The difficulty lies in the fact that the selection of input variables such as test materials and test loads directly affects the variation characteristics of friction information and output variables and tests the real-time collection ability of the data collection system. Therefore, designing a reasonable and reliable test plan and building an ideal and efficient data collection system are the basic requirements for friction and wear tests. This chapter first forms a real-time data collection system for multi-dimensional friction information based on the friction and wear testing machine and combines various test data collection methods. Then, by selecting test pieces and test parameters, preliminary research is conducted on the characteristics of friction information changes. Finally, a test group with load as the control variable is designed to achieve complete collection of raw data for friction and wear testing.

4.3.2.2 Friction test system and friction information

Friction and wear testing is an important method for studying friction processes. The testing system is influenced by set variables such as load, material, and environment. These variables generally do not change during the testing process, but different variables can have an impact on the testing process and results.

The key data in the friction test system are friction-induced products and output. Friction-induced products, also known as friction information, are

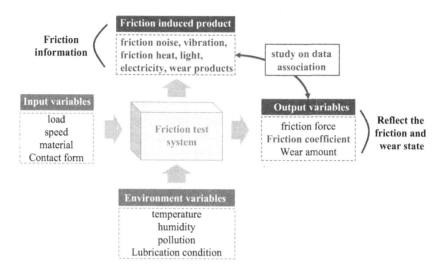

FIGURE 4.71 Friction test system and friction information.

information data generated or induced during the friction and wear process, which involves friction noise, vibration, images, and other friction information. They usually change in real-time with the experimental process and are closely related to output quantities such as friction force and friction coefficient. The friction test system is shown in Figure 4.71.

Friction information is an important product in the process of friction and wear, displaying the changing characteristics of the friction process through various forms. Various friction information has different data characteristics, and real-time data collection needs to be combined with different collection methods. The acoustic, vibration, and image information during the friction process are several intuitive and obvious friction information uses, but the difficulty of real-time collection of different friction information varies greatly, posing challenges to the collection methods.

During the friction process, due to changes in lubrication status or surface morphology, acoustic information such as friction squeal noise and acoustic emission signals may be generated at the friction contact location. Friction squeal noise is closely related to the state of friction and wear. When the lubrication condition is good, the friction noise between metal friction pairs is usually weak. In the dry friction state, the impact vibration and other phenomena caused by surface morphology changes can induce friction squeal noise. In addition, the frictional squeal noise of different materials also exhibits different manifestations. By using sound pressure sensors, digital signal microphones, and other methods, real-time information such as sound pressure and audio frequency of friction squeal noise can be obtained, greatly facilitating the recording, reproduction, and research of acoustic signals during the friction process. And acoustic emission signals are often generated by the rapid release of transient elastic waves on the surface of materials, and their signal characteristics and research methods are similar to vibration signals.

During the friction test process, due to mechanical operation, changes in contact surface roughness, etc., complex vibration information will be generated between the sample and the contact surface, which directly reflects the smoothness of the friction and wear process and is an important characterization of the friction and wear state. The collection methods of vibration information are diverse and can be divided into contact and non-contact methods. The contact type vibration sensor has accurate and reliable data measurement and is favored in mechanical experiments. However, due to the fact that the contact type vibration sensor needs to be installed on the surface of the upper or lower sample, which will affect the friction test conditions, and the friction and wear testing machine used has no magnetized coating on the surface, which cannot directly adsorb the sensor. Therefore, this chapter prioritizes the non-contact collection method. For the vibration information of the sample during the friction process, non-contact measurement methods such as eddy current sensors and laser displacement sensors can be used, which have little impact on the friction test conditions and can effectively obtain vibration information of fixed targets.

Friction and wear occur on the contact surface of materials in relative motion, and surface morphology changes have received high attention in friction tests. Precision instruments such as white light interferometers can capture surface morphology while the friction and wear test are paused, as shown in Figure 4.72. White light interferometers can record three-dimensional information of surface morphology within a specified range.

However, the above methods require pausing or even moving the test specimens, making it difficult to collect them in real time and failing to achieve real-time acquisition and effective utilization of image information during friction and wear processes. In addition to surface morphology, image information during the friction process can also be obtained by observing changes in sample displacement and other methods to characterize the stability of the running process. With the development of computer vision and image processing

(a) Example of the lower sample surface

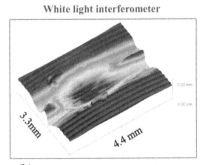

(b) Example of the surface morphology

FIGURE 4.72 Surface topography captured by white light interferometer.

methods, effective acquisition of indirect image information during friction and wear testing is gradually becoming a reality through methods such as motion amplification algorithms.

4.3.2.3 Real-time collection method for multi-dimensional friction information

In order to achieve real-time collection of multi-dimensional friction information in the friction test system, this chapter optimizes the collection method of friction information from the perspective of data collection and processing collaboration and combines various sensing devices to form a real-time collection system of friction information with multi-dimensional data collection, integration, and storage functions.

4.3.2.3.1 Principle of friction information collection test

The various friction information generated in the friction test system has diverse collection methods and varying difficulties in real-time collection. This chapter proposes a schematic diagram of the friction information collection system shown in Figure 4.73, forming a practical and effective real-time collection scheme for friction information, achieving real-time collection of signals such as sound, mechanics, vibration, and images during the friction process.

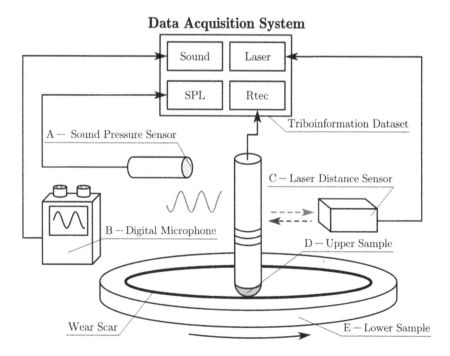

FIGURE 4.73 Schematic diagram of friction information collection system.

Among them, the friction and wear test is conducted in the form of ball disc rotation, with the upper sample being a sphere and the lower sample being a disc. During the test, the lower sample is driven by an electric motor to rotate at a uniform speed, forming friction with the fixed upper sample.

The upper sample is fixed during the test process, facilitating the collection of lateral force and up and down runout data of the upper sample during the friction process through the force sensor and displacement sensor connected to the upper sample. This lateral force is the data source for friction force and friction coefficient. During the jumping process, not only can the load changes be collected, but also the height changes of the upper specimen can be collected, effectively obtaining relevant data such as the contact shape of the contact pair and the stability of the wear state.

For the sound signal during the friction process, the sound pressure sensor and digital microphone are used to collect it separately. The sound signal diffuses around the source of sound (i.e., the contact position) in all directions. Therefore, the collection of acoustic signals is non-contact. By reasonably arranging the sensor positions, signal occlusion can be avoided, and clearer experimental data can be obtained. Two types of sensors can respectively obtain sound pressure and audio data, which can be reproduced through playback, making it easier for the human ear to distinguish changes in the signal. The sound pressure signal stores the air vibration information excited by the friction process through digital signals, with high sensitivity and more primitive data, which helps to save more effective information.

In response to the vibration phenomenon during the friction process, a laser displacement sensor is used to align the laser beam with the center of the upper sample fixture. Through the time difference reflected by the laser beam, the radial vibration displacement data of the upper sample is obtained, which is closely related to the changes in friction force. To ensure the stable and accurate position of the laser beam, it is necessary to avoid interference from the movement of the test bench, build a stable sensor fixture, and adjust the sensor position to ensure that the upper sample fixture is within the sensing range of the sensor.

Due to the difficulty in real-time acquisition of surface morphology, in order to obtain effective image information during the friction process in a reasonable manner, this chapter takes a new approach from the perspective of indirect acquisition of image information, avoiding real-time photography of the friction contact surface morphology. Instead, it focuses on capturing specimens with motion characteristics, using an industrial camera to align with the upper specimen fixture and capture real-time experimental videos, and visual processing methods are used to obtain the motion information of the upper sample from the image information.

The above multiple methods for collecting friction information can obtain real-time friction sound, vibration, and image information, which is conducive to subsequent data processing and analysis. The data are summarized and stored through computers and other devices, forming a complete real-time collection scheme for friction information.

4.3.2.3.2 Real-time collection system for friction information

Based on the schematic diagram, this chapter establishes a real-time collection system for multi-dimensional friction information. Among them, the measurement schemes for mechanical and acoustic data are relatively mature, and the specific situation, layout method, and measurement scheme of the sensing equipment used are as follows:

1. Rtec-5000S friction and wear testing machine

This case uses the Rtec-5000S friction and wear testing machine (hereinafter referred to as the Rtec testing machine) as the testing platform, and the testing machine and its data collection system are shown in Figure 4.74.

The Rtec testing machine has multiple sensors that can achieve real-time collection of test data such as friction force, friction coefficient, upper sample height, and loading force. Among them, the friction coefficient is the target value for the friction and wear test in this chapter, and its value comes from the calculation results of friction data in the testing machine. The height of the upper sample is interrelated with the value of the loading force: The loading force itself is a set value, and to ensure a constant loading force, the height of the upper sample will continuously decrease with increasing wear. The change in the

FIGURE 4.74 Rtec-5000S friction and wear testing machine and data acquisition system.

height of the upper test and the effect of the friction surface also cause the left and right oscillation and up and down oscillation of the upper sample fixture. Therefore, the loading force exhibits variation characteristics during the test process, reflecting the changes in friction and wear status, and the height information of the upper sample also reflects the variation characteristics of wear degree to a certain extent. The sampling frequency of the above data is all 1 kHz, and the experimental data will be preliminarily processed by taking the average value every 10 seconds, that is, down sampling to 100 Hz and automatically saving.

2. GRAS sound pressure sensor

The change in air pressure caused by sound propagation through air is a visual representation of sound signals. This chapter selects the GRAS-46AD sound pressure sensor to collect sound pressure signals inside the experimental cabin. The sound pressure sensor is arranged diagonally above the contact position, with no obstructed object between it and the contact position, a distance of about 10 cm, a sampling frequency of 1 kHz, and a sensitivity of 50 mV/Pa. To achieve automated data collection, the sound pressure sensor information is amplified by the constant current adapter data and transmitted to the upper computer through a data acquisition card, achieving signal reading, collection, and storage.

3. Digital microphone

Audio signals are another manifestation of sound signals, stored and saved in the form of digital signals. During the experiment, a digital microphone was used to collect audio signals inside the test cabin. The digital microphone is placed on the test bench, approximately 10 cm away from the contact position, and the sampling frequency is 48 kHz.

4. Kearns laser displacement sensor

Using a laser displacement sensor to collect radial vibration information of the upper sample, the effective distance range is 80 ± 2 mm, the sampling frequency is 1 kHz, and the horizontal distance between the laser beam and the lower sample is about 5 cm, which does not change throughout the entire experimental process. To achieve automated data collection, laser displacement sensing information is also transmitted to the upper computer through a data acquisition card, achieving signal reading, collection, and storage.

The above sensors and related data collection devices are shown in Figure 4.75, and the collection methods and parameters of each sensor device are shown in Table 4.19.

As mentioned earlier, the sensing equipment needs to fully avoid the movement interference of friction components and be stably installed inside the cabin of the testing machine. Due to the limited space inside the testing

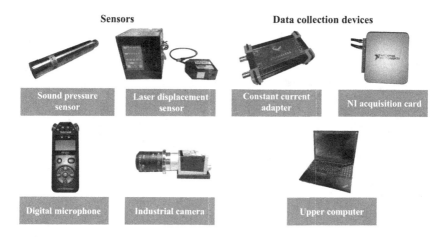

FIGURE 4.75   Data collection equipment.

TABLE 4.19
Data Collection Equipment

Types of Raw Data	Instrument and Equipment	Sampling Frequency (kHz)	Remarks
Friction coefficient	Data collection system of Rtec-	0.1	
Height of upper sample	5000S friction and wear testing machine		/
Loading force			
Sound pressure signal	GRAS sound pressure signal sensor	1	Sensitivity: 50 mV/Pa
Audio signal	Digital microphone	48	/
Vibration signal of upper sample	Keyence laser displacement sensor	1	Effective distance: 80±2 mm
Image information	Industrial camera	/	32 frame/second

machine, a metal support rod is used to build a support and clamping structure, forming a three-dimensional arrangement of the collection system, and installing each sensing equipment. Fix the above sensing equipment on the test bench and bracket in sequence, and complete the layout and installation of the data system collection equipment. The completed real-time data collection system is shown in Figure 4.76.

4.3.2.3.3   Motion amplification algorithm and image acquisition
In friction information, the image data is two-dimensional information, and industrial cameras are used to collect the original data in real time. Under different layout schemes and collection forms, images with different content and

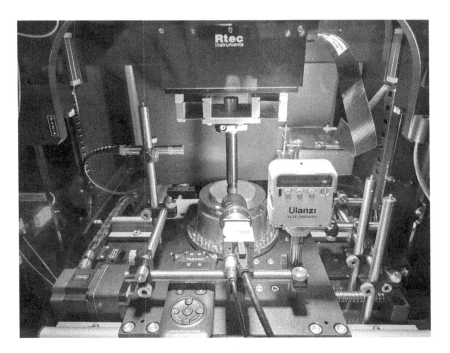

FIGURE 4.76 Test data real-time acquisition system.

data features will be obtained. In the experiment, the industrial camera is placed directly in front of the sample fixture on the testing machine, and the camera axis is parallel to the contact surface of the friction pair to obtain image information of the radial vibration phenomenon of the upper sample. In terms of hardware, an industrial camera is used to capture the corresponding black-and-white image sequence, which is transmitted in real time to the upper computer and saved and converted into video. The actual frame rate is 32 frames per second.

As mentioned earlier, the laser displacement sensor also monitored the radial vibration displacement of the upper sample, and its data were compared with the method of obtaining vibration information from the image. The data information measured by the two methods has a substitution relationship. Therefore, this chapter will use the two types of data separately in subsequent data analysis and use the laser displacement sensor data to verify the effectiveness and accuracy of the image processing method in obtaining information.

In the original image information collected by industrial cameras, the vibration of the upper sample is relatively weak. In order to effectively extract data features from the image information, this chapter uses a motion amplification algorithm to further process the image. This not only makes the motion situation in the image clearer and distinguishable, but also effectively improves the pertinence of image data collection, thereby obtaining image information and data with research value during the experimental process. This provides optimization ideas for the collection method of two-dimensional image data from the perspective of image processing.

The motion amplification algorithm is an image processing method that amplifies small movements in an image. Set the input image data as $I(x, t)$. Observe the position and temporal state of the image, represented by Equation 4.76.

$$I(x, t) = f(x + \delta(x, t)) \tag{4.76}$$

In the equation, $\delta(x, t)$ is the motion area to be amplified, which is related to position x and time t. If the motion amplification coefficient is set to α, output image after motion amplification can be expressed as Equation 4.77.

$$\tilde{I}(x, t) = f(x + (1 + \alpha)\delta(x, t)) \tag{4.77}$$

In practice, it is usually only necessary to zoom in on specific motion regions of interest. Therefore, it is necessary to select the motion regions in the image, and the selector can be expressed as Equation 4.78:

$$\tilde{\delta}(x, t) = \sigma(\delta(x, t)) \tag{4.78}$$

In the equation, σ is a temporal bandpass filter.

The time domain bandpass filter of traditional motion amplification algorithms generally requires manual design and is susceptible to phenomena such as object occlusion and image noise in the image, resulting in poor filtering performance. The motion amplification algorithm based on deep learning can be used to learn filters from multiple actual object datasets, thereby obtaining spatial decomposition filters with better amplification effects. It can separate foreground and background in images and has strong generalization ability.

The process of motion amplification algorithm based on deep learning is shown in Figure 4.77, consisting of main modules such as spatial decomposition

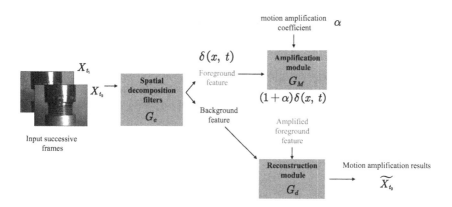

FIGURE 4.77 Motion magnification algorithm flow.

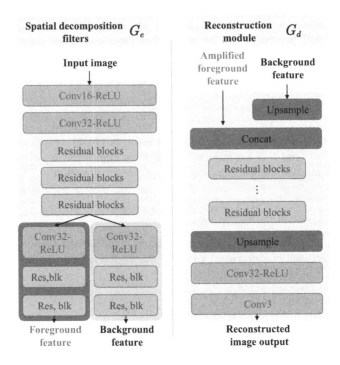

FIGURE 4.78 Details of motion magnification algorithm.

filter, amplification module, and reconstruction module. After separating the foreground and background through a spatial decomposition filter, an amplification module is used to amplify the foreground motion features. Finally, a reconstruction module is used to reintegrate the enlarged foreground and background features to obtain motion amplification results. Video results can also be obtained by synthesizing video sequences.

Among them, the spatial decomposition filter and reconstruction module are neural networks composed of convolution module (Conv), upsample module (Upsample), concat module (Concat), residual blocks (Res blk), and other modules. Their internal structure is shown in Figure 4.78.

The amplification module uses a linear amplification method to amplify the foreground motion features of two frames of images. Set two frames of image X_a and X_b. After passing through spatial decomposition filters, the foreground feature obtained is M_a and M_b. The operational equation of the amplification module can be expressed as Equation 4.79.

$$G_m (M_a + M_b, \alpha) = M_a + \alpha (M_b - M_a) \tag{4.79}$$

Due to the small size of the motion of interest, using traditional motion amplification algorithms to process images can easily generate noise or excessive blur, which can have a negative impact on image processing. The motion

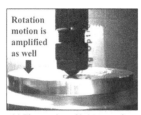

(a) The rotation of lower sample affects analysis results

(b) Brightness affects edge feature extraction

(c) Distinct edge feature

FIGURE 4.79 Image requirements for motion magnification algorithm.

amplification algorithm based on deep learning effectively avoids the interference of image noise by separating the front and back scenes, but non-focused large amplitude movements can also cause interference to the algorithm. Especially in scenarios where multiple movements overlap, the enlarged image may weaken the clarity of the motion of interest.

In the process of image acquisition, selecting the appropriate shooting angle to avoid interference from corresponding objects or movements is the most direct and effective improvement method. As shown in Figure 4.79, there are two main interferences in the collection and analysis process of friction and wear test image data: (1) the rotational motion of the lower sample is significantly greater than the vibration amplitude of the upper sample fixture, and (2) edge feature blurring is caused by supplementary lighting.

Therefore, the above interference factors pose requirements for image acquisition from two perspectives: (1) try to obtain a clean foreground and background environment, and (2) focus on the area of interest. As shown in Figure 4.79, by improving the brightness and background environment, increasing the magnification lens, and other methods to enlarge the area of interest, a clear and obvious shooting angle with edge features was gradually obtained. Combined with friction and wear tests, image data collection was completed.

4.3.2.4 Friction and wear test and data acquisition
On the basis of a multi-dimensional friction information real-time collection system, scientific methods are used to select test materials, sct up tcst plans, and conduct systematic friction and wear tests, in order to obtain rigorous and reliable multi-dimensional friction information test data, providing convenience for subsequent data processing and analysis.

4.3.2.4.1 Selection of test conditions
Generally, tribological behavior is closely related to test conditions such as friction pair materials and test duration. The characteristics of friction and wear data that can be observed and collected vary depending on the combination of different test conditions. Before the experiment, by combining various experimental parameters, a preliminary understanding of friction and wear behavior

and variation patterns can provide effective reference for the design of the experimental plan. There is a close relationship between the contact material and the available friction information, and parameters such as test duration have a significant impact on the data collection plan and data processing. Therefore, the pretest is carried out from two perspectives: the contact material and the test duration.

In order to ensure that the experimental group can obtain sound, vibration, and image information with changing characteristics and obvious phenomena, this chapter designs two sets of pretest plans to observe and analyze the friction and wear test process under different materials and different conditions of different lengths, and form a preliminary understanding of the friction and wear process, friction phenomena, and changes in friction information under different materials and conditions of different lengths during the test process, and provide materials and parameter benchmarks for formal testing.

In the pretest, the friction and wear test adopt a ball disc contact form, with the upper sample sphere having a diameter of 9.525 millimeters and the lower sample disc having a diameter of 70 millimeters and a thickness of 6 millimeters. Before each group of tests, new samples are replaced, and the test duration, load, and other parameters are set using a testing machine.

4.3.2.4.1.1 Pretesting of different contact materials   In the pretest of friction and wear contact pair materials, in order to control variables, GCr15 spheres were selected for the upper samples, and discs made of cast iron, copper, and aluminum alloys were selected for the lower samples. The test duration was 30 minutes, and the load was $10N$. During the test, friction noise, vibration, and other phenomena were observed and recorded throughout the entire process. The specific test plan is shown in Table 4.20.

Through preliminary analysis of pretest data under different contact pair materials, it is shown that there are significant differences in friction behavior between different material pairs: When the lower sample material is cast iron, the friction noise is very small and easily masked by environmental noise. When the lower sample material is copper, the friction noise is also relatively weak, and the evolution process of friction and wear is stable, with no obvious trend of change. When the lower sample is made of aluminum, the phenomenon of frictional squeal noise is significant, with a constantly increasing trend of

TABLE 4.20

Pretest Plan for Different Contact Pair Materials

Material of Upper Sample	Material of Lower Sample	Duration (min)
GCr15	Cast iron	30
GCr15	Copper	30
GCr15	Aluminum alloys	30

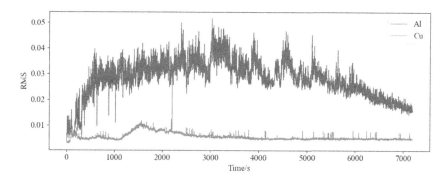

FIGURE 4.80   Comparison of friction noise data for different materials.

variation, and accompanied by more obvious vibration phenomena, the monitored friction coefficient and other numerical fluctuations are intensified.

Figure 4.80 shows the entire experimental process, and the root mean square data of friction noise over time under two material conditions. Among them, under the condition of aluminum alloy samples, the root mean square value of sound pressure throughout the entire process is much greater than that of copper samples, and the peak data of the former is about 4 times that of the latter, which effectively confirms the differences in acoustic characteristics distinguished by auditory perception. However, the drawback is that under the aluminum alloy sample, the sound pressure signal curve fluctuates significantly, indicating that the stability of the acoustic data collected during the test process is not as good as that of the copper sample.

The micro morphology is an important reason for the differences in acoustic characteristics mentioned above. Furthermore, a white light interferometer was used to observe the surface morphology of the three materials in greater detail, as shown in Figure 4.81. The results show that the surface wear marks of the cast iron lower sample are extremely shallow, making it difficult to distinguish from the non-worn area. The surface wear marks of the copper sample are very uniform, forming a good distinction from the non-worn area. The depth of the wear marks is about 8 micrometers, and the width is relatively narrow, about 0.5 millimeters. Compared with the two samples mentioned above, deeper and wider wear marks were formed on the surface of the aluminum alloy lower sample, with a depth of about 78 micrometers and a width of about 1.4 millimeters. The center of the wear marks had irregular protrusions, which were residual materials during the wear process. It is precisely due to the non-uniformity of wear marks that aluminum alloy specimens generate more obvious friction noise during the testing process.

From the perspective of friction information research, a combination of test materials with more obvious phenomena such as noise and vibration should be selected in order to obtain more effective friction information test data. At the same time, achieving effective monitoring of the relatively unstable friction and wear test process has more practical value.

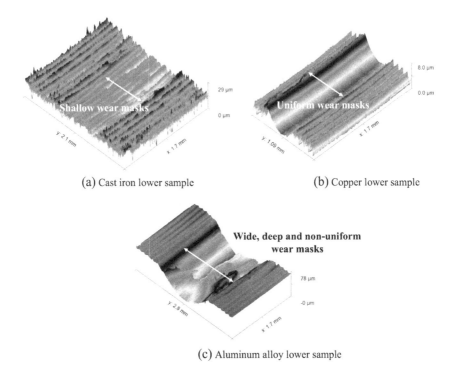

(a) Cast iron lower sample   (b) Copper lower sample

(c) Aluminum alloy lower sample

FIGURE 4.81   Comparison of worn surface morphologies of different materials.

4.3.2.4.1.2 Pretesting with different test durations   In the pretest for the duration of friction and wear tests, both GCr15 upper and aluminum alloy lower specimens were used as the test materials, with a load of $10N$. The test duration was controlled for 10, 30, and 120 minutes, respectively. The changes in friction coefficient and friction information data during the test process were observed. The specific test plan is shown in Table 4.21.

The experimental results are shown in Figure 4.82, recording the variation of the coefficient of friction (COF) over time under three different time conditions. The results indicate that there are differences in redundant information in long-term friction tests at different times.

TABLE 4.21
Pretest Schemes for Different Test Durations

Material of Upper Sample	Material of Lower Sample	Duration (min)
GCr15	Aluminum alloys	10
GCr15	Aluminum alloys	30
GCr15	Aluminum alloys	120

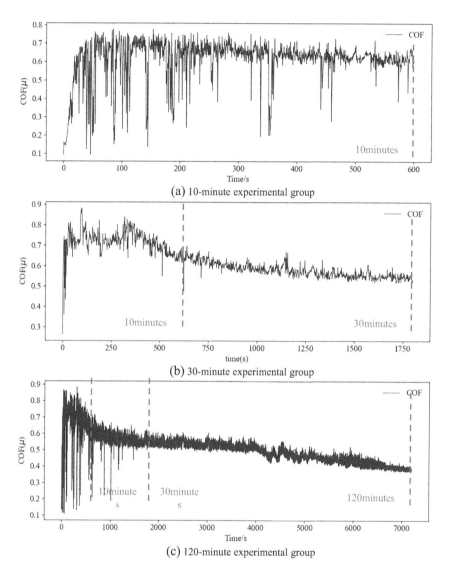

(a) 10-minute experimental group

(b) 30-minute experimental group

(c) 120-minute experimental group

FIGURE 4.82 Comparison of friction coefficient changes in different time tests.

Friction and wear is a continuous process, and long-term friction and wear tests basically contain all the information and characteristics of short-term friction and wear tests. At the same time, the phenomenon of friction and wear varies at different stages of the test. Taking the 120-minute test data as an example, in the early stage of the test process, the friction and wear change quickly, and the phenomenon is obvious. After 10 to 30 minutes, it gradually turns into a stable state, and the changes in friction coefficient and various friction information are significantly weakened. After 30 minutes, the redundant information in the friction and wear test increases significantly, and the observable change characteristics are extremely unclear.

From the perspective of real-time monitoring and analysis of data, it is necessary to fully collect experimental data with rich information and obvious features, and avoid the concealment of valuable data by redundant information. Prolonged data collection can also increase the burden of data processing if it fails to provide information that is conducive to the study of related features.

In summary, this chapter selects aluminum alloy as the lower sample material, sets the test duration to 30 minutes, designs a formal test plan with load as the control variable, and implements the test plan using standard test procedures to collect complete test data.

4.3.2.4.2   Test samples

The friction and wear tests were conducted using a ball disc contact method. The upper sample was made of GCr15 metal balls with a diameter of 9.525 millimeters, while the lower sample was made of aluminum alloy discs (grade 6061) with a diameter of 70 millimeters and a thickness of 6 millimeters. New samples were used for each group of tests, and the surface oil stains were cleaned using an ultrasonic cleaner before the test. The sample material is shown in Figure 4.83.

4.3.2.4.3   Control variable test

In practical production and life, the friction conditions are complex and variable, so multiple working conditions are often designed for the friction and wear test process to obtain more diverse and valuable data results. In order to obtain experimental data results under multiple operating conditions and enhance the universality and representativeness of data collection and analysis, this chapter designs and conducts control variable experiments using load as the experimental variable. In order to collect clear image data of the friction test process

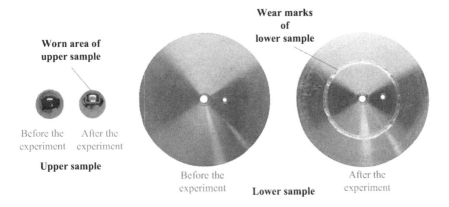

FIGURE 4.83   Schematic diagram of test materials.

and avoid further interference with the sound signal, the formal tests in this chapter are all metal-to-metal dry friction tests without lubricating oil or coolant.

First, from the perspective of friction noise analysis and comparison, an experimental group and an environmental sound control group were set up, and specific plans were described separately.

4.3.2.4.3.1 Experimental group   Set up different axial load test groups, repeat the test twice for each parameter combination, and complete data collection separately. The experimental parameters are shown in Table 4.22, where the rotational speed information represents the number of revolutions per minute (rpm) of the sample. All parameters can be set through the testing machine before the test.

4.3.2.4.3.2 Environmental sound control group   Set up three sets of environmental sound control groups and collect environmental sound in the test cabin for three situations: The test machine is turned off, the test machine is turned on but not running, and the test machine is turned on and running without load. The specific test parameters are shown in Table 4.23.

Second, to ensure the reliability and reference value of the pretest data, the formal tests are strictly carried out in accordance with the standard operating procedures. The specific test steps are as follows:

TABLE 4.22
Test Group Plan

No.	Rotational Speed/rpm	Radius of Rotation/mm	Duration/min	Load/N
1	60	20	30	20
2	60	20	30	10
3	60	20	30	5

TABLE 4.23
Ambient Sound Control Program

No.	Operation Condition of Testing Machine		Duration/min	Contact Condition
	Rotational Speed/rpm	Radius of Rotation/mm		
4	60	20	5	Non-contact
5	Testing machine is turned on but not running		5	Non-contact
6	Testing machine turned off		5	Non-contact

1. **Sample preparation:** Each group of experiments uses brand new upper and lower samples. Before the experiment, the surface of the samples is cleaned with anhydrous ethanol ultrasound for 30 minutes for oil stains. After cleaning, the surface is wiped dry with a non-woven cloth.

2. **Sample installation:** Use fixtures to install the upper and lower samples separately, where the upper sample fixture only retains vertical degrees of freedom for adjusting the sample height and loading force. The lower sample retains rotational freedom, and during the test process, the ball disc rotation friction and wear experiment will be completed through the rotation of the lower sample.

3. **Testing and data collection:** Turn on the digital microphone, upper computer data collection program, testing machine, and industrial camera in sequence, keep the test cabin door closed, and collect complete test data for the entire friction and wear process. The testing machine first adjusts the load to the preset value, and after holding the load for one minute, the motor of the testing machine will drive the lower sample to rotate, completing the predetermined duration of ball disc rotation friction and wear test. After the experiment, turn off the testing machine, industrial camera, upper computer data acquisition program, and digital microphone in sequence to complete data acquisition and save.

4.3.2.5 Summary

This chapter is based on the friction and wear testing machine and optimizes the multi-dimensional friction information collection method by combining data processing methods. A real-time data collection system for friction and wear testing has been established, achieving effective collection of multi-dimensional friction information raw data. Preliminary experiments were conducted using a combination of multiple materials, multiple durations, and other experimental conditions to enhance the data collection effect of friction information and reduce information redundancy. This provides a reference for the material and parameter design of formal experiments. Subsequently, multiple working conditions, repeatable control variable formal experiments, and environmental sound control experiments were designed and implemented, forming a rigorous and reliable experimental combination. Rich and complete multi-dimensional friction information raw data were obtained, laying the foundation for the subsequent research on the correlation characteristics of friction information and friction coefficient.

4.3.3 FEATURE EXTRACTION AND FUSION OF MULTI-DIMENSIONAL FRICTION INFORMATION

4.3.3.1 Introduction

The original data of friction and wear tests are influenced by environmental factors, collection methods, and other factors, which inevitably lead to issues such as data noise and data redundancy. Their effective information is easily disturbed and hidden by data errors. Therefore, it is necessary to extract effective information highlighting the characteristics of friction and wear changes from the original data through feature engineering methods. From experiments, multi-dimensional friction information such as acoustics, vibration, and images have been obtained during the dry friction process. This chapter organically combines the data features of various types of friction information with the changes in the dry friction process. Through feature selection and feature extraction, feature data with tribological research value are excavated and extracted from the original data. Subsequently, the feature data are segmented and aligned using the time slicing method to achieve effective fusion of data from different dimensions and sources, and a preliminary analysis is conducted on the changing trend of multi-dimensional friction information.

4.3.3.2 Data feature extraction of one-dimensional friction information

There is various one-dimensional friction information in the friction and wear test data, with high sampling frequency and susceptibility to noise interference. Therefore, digital signal processing methods such as time-frequency domain analysis should be used to perform feature analysis on the original signal, and statistical methods should be used to achieve data downsampling to improve the stability and effectiveness of the feature data. This section discusses different friction information data processing methods based on different collection methods and fully combines the characteristics of the data with the changes in the dry friction process to achieve friction information feature extraction with tribological research value.

4.3.3.2.1 Testing machine data

The testing machine collected various data such as friction coefficient, loading force, and height of the upper sample. The original data had a high sampling frequency and significant data fluctuations. Effective information should be extracted through data processing methods such as filtering to highlight the trends and characteristics of each data change.

4.3.3.2.1.1 Friction coefficient For the target value friction coefficient, a low-pass filtering method based on Butterworth filter design is used to filter the data, effectively highlighting the trend of friction coefficient changes and enhancing the generalization ability of data analysis. Figure 4.84 shows the curve of the filtered friction coefficient over time.

Under different load conditions, the friction coefficient curve generally shows a clear trend of increasing rapidly at first and then gradually decreasing and can be roughly divided into two stages: fluctuation stage and stable stage. The fluctuation stage is mainly located in the early stage of friction coefficient change, and as the load increases, the data fluctuation gradually weakens. Under low load ($5N$) conditions, the data fluctuation is the most severe. The test lasted for 10–15 minutes, and the friction coefficient under each load generally entered

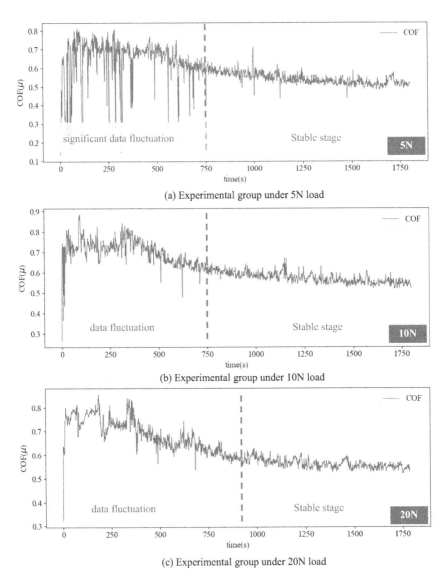

(a) Experimental group under 5N load

(b) Experimental group under 10N load

(c) Experimental group under 20N load

FIGURE 4.84 Variation curve of friction coefficient with time.

a stable period of weakening and slowly decreasing fluctuations and continued until the end of the test.

The variation curve of friction coefficient indicates that the characteristic of first increasing and then decreasing friction coefficient is universal, indicating the effectiveness of experimental data. At the same time, it still has a lot of volatility and randomness, which is a difficulty, and focuses on subsequent data processing and research. By the end of the experiment, the friction coefficient had been in a relatively stable state, and the test data had collected various changes in characteristics such as fluctuations, fluctuations, and stability, which was conducive to forming a rich and representative friction test dataset.

The friction coefficient exhibits the above changing characteristics, which are closely related to the friction behavior mechanism and the material properties of the friction pair. In the initial stage of friction and wear tests, the dense oxide film on the surface of aluminum alloy participates in friction. Increasing the load helps to quickly wear the oxide film, thereby avoiding severe fluctuations in friction coefficient. The load affects the variation of friction force through the size and deformation state of the contact area. As time goes on, the increasing trend of the contact area gradually stabilizes; therefore, the change in friction coefficient enters a relatively stable stage.

4.3.3.2.1.2 Loading force   The loading force itself is the test set value. However, during the experiment, to ensure a constant loading force, the upper sample will undergo longitudinal vibration with the surface morphology, resulting in irregular fluctuations in the loading force data. As shown in Figure 4.85, the filtered loading force data (Fz_mean) remains stable near the set value, indicating that the overall loading force during the test process is stable, but it cannot effectively reflect the fluctuation characteristics of the test process.

In the figure, the orange curve (labeled Fz_min) is sampled from the lower envelope of the loading force. As time increases, the fluctuation of the lower envelope data of the loading force decreases, indicating that the fluctuation of the loading force becomes more severe. This indicates that compared to the average value of the loading force, its lower envelope line can more clearly reflect the increasing trend of loading force fluctuations. Therefore, in order to reflect the changing characteristics of the loading force more effectively, the lower envelope value (Fz_min) of the loading force is selected as the feature of the loading force data.

4.3.3.2.1.3 Height of upper sample   During the friction and wear test, as the degree of wear deepens, the height of the upper sample will gradually decrease. Therefore, the height data of the upper sample is to some extent related to the wear phenomenon during the test process. During the test, the testing machine is equipped with a sensing device for the height of the upper sample.

As shown in Figure 4.86, the height data of the filtered upper sample is closely related to the trend of friction coefficient change, with a rapid increase in the early stage and a gradual decrease. The data themselves show obvious changes. Therefore, after taking the average value of the data at intervals of seconds, the average value of the upper test height (marked as Z_mean in the figure) is taken as the data feature.

4.3.3.2.2   Sound pressure sensor data

The sound pressure sensor data are amplified by a signal exciter and stored as a digital signal. First, the time-frequency spectrum method is used to preliminarily identify the noise characteristics of the data.

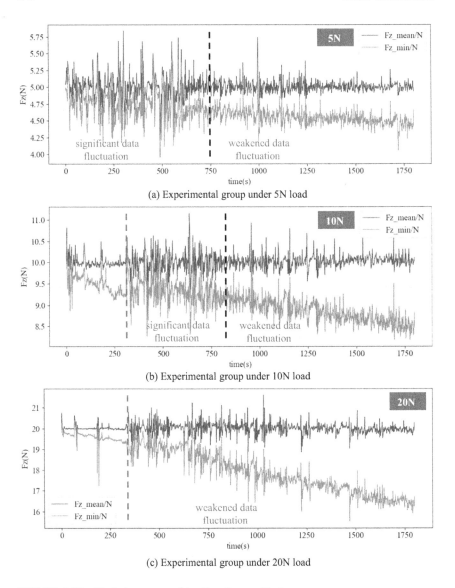

(a) Experimental group under 5N load

(b) Experimental group under 10N load

(c) Experimental group under 20N load

FIGURE 4.85   Variation curve of loading force with time.

Based on the short-term Fourier transform (STFT) method, time-frequency domain analysis was conducted on the environmental sound data and formal test group sound pressure data of the testing machine during no-load operation. The obtained time-frequency spectrum is shown in Figure 4.87, and each image consists of two parts: (1) the time-frequency spectrum is located in the upper right corner of each image, with the horizontal axis representing the time axis and the vertical axis representing the frequency value. The darker the color, the

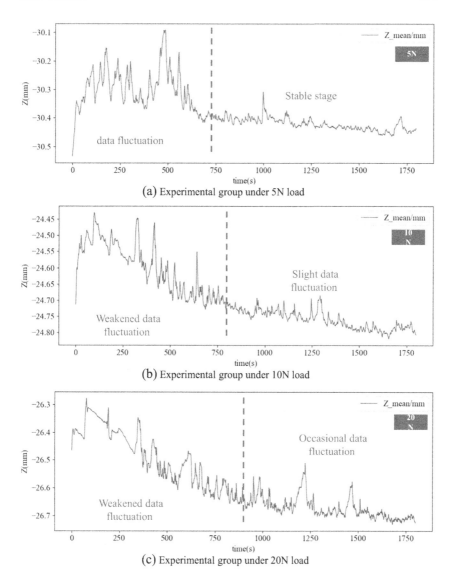

(a) Experimental group under 5N load

(b) Experimental group under 10N load

(c) Experimental group under 20N load

FIGURE 4.86   The height of the upper sample varies with time.

higher the amplitude at the corresponding frequency at that time; (2) For different load conditions, except for no load, the frequency spectrum is plotted at the same time (500 s) during the test process. The horizontal axis in the graph represents the frequency value, and the vertical axis represents the frequency amplitude. The higher the amplitude, the stronger the frequency signal at that time.

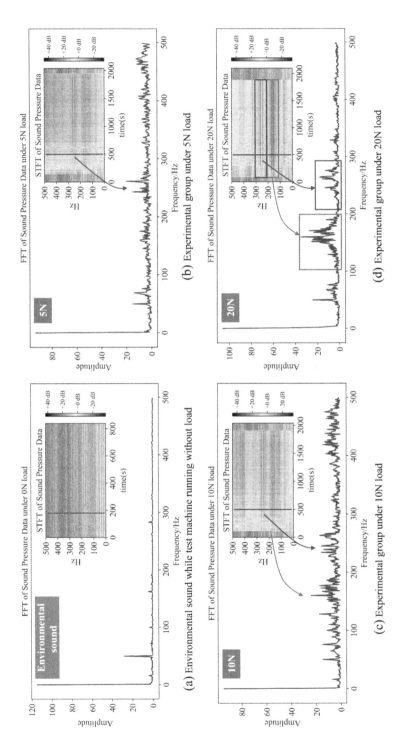

FIGURE 4.87 Frequency spectrum diagram of noise.

The spectrum diagram shows that both environmental noise and sound pressure data from the formal testing process have obvious frequency band distribution characteristics. The environmental noise collected during the no-load operation test (test number 6) is mainly distributed in the low-frequency range of 0–100 Hz. As the load increases, the experimental sound pressure data show increasingly obvious frequency changes in the frequency bands of 100–200 Hz and 200–300 Hz, indicating that during the friction and wear test process, the energy of the sound pressure signal gradually concentrates towards the above frequency bands; the other frequency bands did not show a significant trend of change.

In addition, during observation, the trend of spectral changes was observed. Before and after the experiment, the colors of each frequency band were relatively light, highlighting the strong correlation between dark regions and the experimental process. During the experiment, the color of the 100–200 Hz and 200–300 Hz frequency bands quickly changed from light to deep and remained stable, forming a sharp contrast with the light-colored areas of other frequency bands, indicating that the sound signal energy in the above frequency bands was concentrated and stable.

In response to the frequency band distribution characteristics generated by the above sound pressure signals, in order to obtain frequency division data with key features in specific frequency bands such as 100–200 Hz and 200–300 Hz, while effectively avoiding data interference from low-frequency noise, a bandpass filtering method based on the Butterworth filter is first used to divide the sound pressure data of each group of experiments. The time-frequency spectra before and after the division are shown in Figure 4.88.

By frequency division, the characteristic information of sound pressure data in the 100–200 Hz and 200–300 Hz frequency bands have been effectively paid attention to, achieving effective separation of corresponding frequency band sound pressure data and avoiding interference from other frequency band noise on data validity.

Furthermore, for the corresponding frequency band of sound pressure data, it is necessary to combine acoustic knowledge for feature extraction to intuitively reflect the temporal trend of sound pressure data. At the same time, in order to ensure the consistency between data changes and human auditory perception as much as possible, acoustic data features with human auditory perception characteristics should be fully selected.

Research has shown that the perception of sound changes by the human ear is nonlinear. In the low-frequency region, the human ear feels more sensitive, while in the high-frequency region, the human ear's perception becomes rougher and rougher. Therefore, in the feature extraction of acoustic signals, full consideration should be given to reproducing the human ear's perception characteristics of sound through mathematical methods.

Acoustic signal research often uses root mean square as the effective value to characterize the energy level of the sound pressure signal. Then, through the logarithmic method, the effective value of the sound pressure is converted into

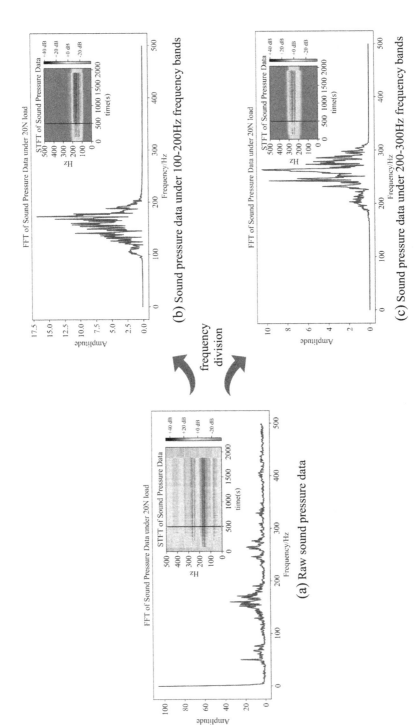

FIGURE 4.88 Spectrum diagram of sound pressure signal before and after frequency division.

SPL features, usually in decibels (dB). The method for calculating root mean square values in the time domain is shown in Equation 4.5.

$$P_{rms} = \sqrt{\frac{1}{k}\sum_{i=1}^{k}p_i^2}, i = 1, 2, \cdots, k \qquad (4.80)$$

In the equation, k is the total number of sample points within the calculation interval, p_i is the sound pressure value of the sample point.

The method of converting the root mean square value to SPL is shown in Equation 4.81.

$$SPL\,(dB) = 10\log_{10}\left(\frac{P_{rms}}{P_{ref}}\right)^2 = 20\log_{10}\left(\frac{P_{rms}}{P_{ref}}\right) \qquad (4.81)$$

$P_{ref} = 2 \times 10^{-5}Pa$. This is the reference value for SPL.

In summary, for sound pressure data under different loads, root mean square values were calculated at second intervals and converted into SPLs. The SPL data features were extracted for the 100–200 Hz frequency band, 200–300 Hz frequency band, and full frequency band, respectively. The curves of their changes over time are shown in Figure 4.89.

The overall trend of SPL data is increasing, and compared with characteristic data such as friction coefficient, it also has obvious fluctuation and segmentation characteristics. The difference is that under low load (5N) conditions, the SPL data fluctuates significantly in the early stage and remains relatively stable in the later stage (about 15 minutes later). Under the conditions of medium load (10N) and high load (20N), the SPL data fluctuated slightly in the early stage, but there were multiple fluctuations. In the later stage, it entered a stable stage quickly, lasted for a long time, and had obvious segmentation characteristics.

In addition, the development trend of the relevant frequency band sound pressure characteristics reflected by the time-frequency spectrum (as shown in Figure 4.87) is more clearly displayed in the data curve. Under low load conditions, the characteristic data of SPLs in the 100–200 Hz and 200–300 Hz frequency bands are almost the same, and the trend of change remains consistent. The spectral line heights in the time-frequency spectrum are also relatively close. Under medium load conditions, the characteristic data of SPLs in the two frequency bands gradually separate. The characteristic data of SPLs in the 100–200 Hz frequency band are slightly higher than those in the 200–300 Hz frequency band, but the trend of change is still basically the same. In the time-frequency spectrum, the spectrum data of the 100–200 Hz frequency band are also significantly enhanced. Under high load conditions, the SPL characteristic data of the 100–200 Hz frequency band and the 200–300 Hz frequency band are significantly separated, and the fluctuation of the 100–200 Hz frequency band is significantly reduced in the later stage. Correspondingly, in the time-frequency spectrum, the spectral line characteristics of the 100–200 Hz frequency band are

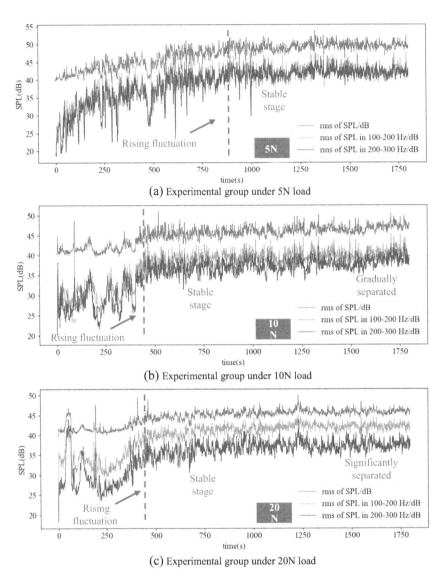

(a) Experimental group under 5N load

(b) Experimental group under 10N load

(c) Experimental group under 20N load

FIGURE 4.89 Curve of sound pressure data over time.

also significantly higher than those of other frequency bands. In summary, the three types of SPL feature data exhibit gradually separated features, but the changing features are highly correlated.

4.3.3.2.3 Audio sensing data

The sound signal collected by the digital microphone is audio data, and root mean square values and Mel-scale frequency cepstral coefficients (MFCC) are extracted as data features.

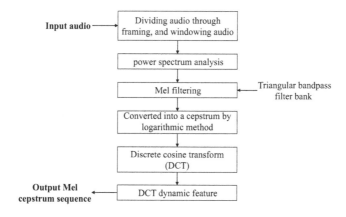

FIGURE 4.90   Mel cepstrum analysis process.

The Mel frequency analysis method simulates human auditory perception from two aspects. First, a nonlinear scale called Mel-scale is defined, which is based on the pitch definition and can be converted from ordinary frequency to Mel frequency using Equation 4.82.

$$mel(f) = 2595 * \log_{10}\left(1 + \frac{f}{700}\right) \tag{4.82}$$

The Mel-scale reflects the degree to which the general human ear perceives frequency. Second, the perception of audio data was focused on specific areas through filter banks, simulating the auditory perception characteristics of humans focusing on partial frequencies.

The above characteristics have made the Mel cepstrum sequence achieve outstanding results in the research of natural language processing problems such as speech recognition and have been widely applied in the resolution of complex sound sources. It has excellent performance in distinguishing different timbres of sound and has reference significance for the feature extraction of audio data in this chapter, especially for highlighting experimental related audio and reducing environmental sound interference. The calculation process is shown in Figure 4.90.

First, the audio is divided into shorter segments through framing, which has better stability compared to longer time segments. Then, calculate the power spectrum for the data and use the Mel filtering method to filter and process the data. The difference between Mel filtering and other filtering methods lies in the use of a triangular overlapping window function to form a bandpass filter bank, which focuses on the center frequency of the triangular window function while converting the ordinary frequency to a more auditory Mel frequency.

As the frequency changes from low to high, the bandwidth of the triangular bandpass filter bank will become wider to collect a larger range of frequency features, which is in line with the rough perception of the human ear in the high-frequency range. At the same time, the filter will also be set from dense to sparse, with bandpass intervals overlapping each other to smooth the spectrum. At the same time, it also plays a role in eliminating harmonics and highlighting audio resonance peaks, reducing computational complexity.

Next, the logarithmic method is used to convert it into a cepstrum. Finally, the discrete cosine transform (DCT) is used to implement the inverse transform and remove the correlation caused by overlapping triangular window functions, obtaining the MFCC. Relatively speaking, the MFCC coefficients in the low-frequency range can bring better recognition characteristics, so it is often necessary to extract some low-frequency range values after DCT transformation as practical MFCC feature data.

The experimental audio data will be divided into small segments at second intervals before input, and input in the form of digital signals. A filter bank of size 128 will be used, and 20 dynamic numerical features of the low-frequency part will be taken after DCT transformation to form an MFCC sequence. Considering the requirement of aligning with other feature data, the MFCC sequence per second is compressed into a single value by taking the average value, ultimately forming a feature value time series with second intervals. The characteristic value variation curves of audio data from each experimental group are shown in Figure 4.91.

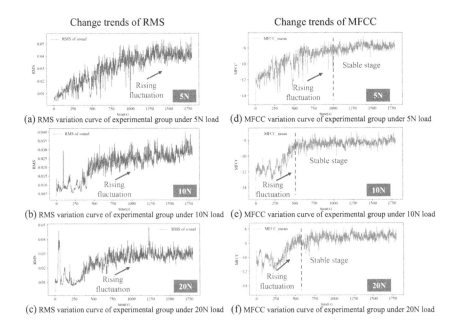

(a) RMS variation curve of experimental group under 5N load (d) MFCC variation curve of experimental group under 5N load

(b) RMS variation curve of experimental group under 10N load (e) MFCC variation curve of experimental group under 10N load

(c) RMS variation curve of experimental group under 20N load (f) MFCC variation curve of experimental group under 20N load

FIGURE 4.91 Audio characteristic data change curve over time.

The figure shows the variation of two acoustic characteristics (root mean square and MFCC) data over time under three different load conditions. Under different load conditions, the root mean square characteristics show a continuous upward trend of fluctuation, but the changing trends of the three are different in the 0–300 s range: under low load conditions, the root mean square characteristic values still have strong short-term fluctuations, while under medium to high load conditions, the short-term fluctuations of the root mean square characteristic values decrease, but there will be obvious upward and downward changes, reflected in the peak changes of the fine and high curves in the figure. Under different load conditions, the MFCC characteristics show a trend of first fluctuating and then maintaining stable data changes. Under medium to high load conditions, the MFCC values reach a stable stage faster, while the fluctuation stage shows a trend of first decreasing and then increasing.

A comprehensive comparison of audio data features and SPL data features can indicate that the root mean square data trend of audio features is highly similar to the SPL data trend and has a strong correlation with the MFCC data trend. Both feature data fluctuate and increase, indicating that the friction noise data generally increases with the friction and wear process, which is basically consistent with the human perception during the experimental process.

4.3.3.2.4 Laser sensor data

The measurement data of laser sensors are displacement signals used to represent vibration characteristics. Common vibration signals can be extracted through various methods such as time domain and frequency domain. This chapter focuses on the displacement signals collected by laser sensors, extracting the kurtosis (Kurtosis, Kurt) and skewness (Skewness, Skew) of signal changes as feature values.

Kurtosis is a statistical feature that measures the sharpness of data distribution, reflecting the steepness of data distribution morphology. It is generally defined based on the fourth-order standard moment of the sample and is actually expressed as Equation 4.83 using a calculation method.

$$kurt(X) = E\left[\left(\frac{X-\mu}{\sigma}\right)^4\right] - 3 \tag{4.83}$$

The term "−3" included in the equation results in a kurtosis value of 0 in a normal distribution. Correspondingly, when the kurtosis value is greater than 0, the data distribution will be steeper than the peak of the normal distribution, indicating a sharp change in value. When the kurtosis value is less than 0, the data distribution will be smoother than the peak of the normal distribution, indicating a relatively small change in value.

Skewness is a statistical feature that measures the direction and degree of skewness in data distribution, reflecting the degree of asymmetry in data distribution. It is defined as the third-order standard moment of the sample, and its calculation method can be expressed as Equation 4.84.

$$skew(X) = E\left[\left(\frac{X-\mu}{\sigma}\right)^3\right] \qquad (4.84)$$

The larger the absolute value of skewness, the greater the degree of deviation in the data distribution. When the skewness value is 0, the data distribution pattern is the same as the skewness of a normal distribution, that is, a symmetric distribution. When the skewness value is greater than 0, the data distribution is positive or right skewed, indicating that the tail of the data distribution is longer at the right end. When the skewness value is less than 0, the data distribution is negative or left skewed, indicating that the tail of the data distribution is longer at the left end.

For the raw data collected by the laser displacement sensor, the data are segmented at second intervals, and the kurtosis and skewness feature values are calculated for each small segment of data to form a feature data sequence that characterizes the lateral vibration data of the upper sample. As shown in Figure 4.92, the variation curves of Skew and Kurt eigenvalues over time under three different load conditions are shown.

Observing the eigenvalues sequence, the kurt value is generally in a positive range, showing a trend of first increasing and then decreasing. The trend remains

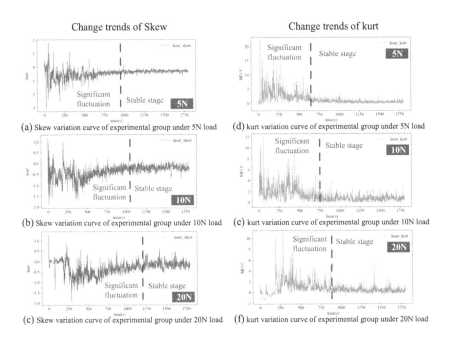

FIGURE 4.92   The data characteristics of the laser sensor change with time.

consistent with the change in friction coefficient. The initial vibration signal is sharp, and the vibration phenomenon is more intense. As friction and wear gradually evolve, the vibration phenomenon gradually tends to stabilize.

The Skew value is generally in a negative range, and its absolute value also shows a trend of increasing first and then decreasing, indicating that the radial vibration phenomenon of the upper sample fixture is not equal amplitude oscillation due to the contact surface effect during the friction process. The unidirectional deviation of the vibration signal in the early stage is more obvious, and as the experiment progresses, the vibration phenomenon gradually transitions to equal amplitude oscillation.

4.4.3.3   Data feature extraction of 2D image information

For the image sequence of the upper sample collected by industrial cameras, a motion amplification algorithm was used to amplify the small movements. The motion frequency remains unchanged, the image clarity slightly decreases, and the motion amplitude becomes more obvious. To further achieve data dimensionality reduction, more comprehensive image processing methods are needed to extract effective information such as friction vibration and achieve scientific processing of two-dimensional friction image data.

4.4.3.3.1   Image feature extraction method

By using the deep learning-based motion amplification algorithm, effective amplification of motion features in the friction test image sequence was achieved. To extract the changing features in the image sequence and form data features that reflect the radial vibration of the upper sample, further image processing and feature extraction are required for the above image sequence, which is divided into six steps, as shown in Figure 4.93.

First, for each frame in the image sequence, focus on the ROI and extract features from the banded data. Among them, the ROI area needs to be selected according to the actual situation. In this chapter, the location is selected in the upper part of the image, which is a thin linear area that runs through the upper sample. In this ROI region, the upper sample portion has a high proportion and multiple edge features with significant color differences, which is beneficial for extracting image features and using statistical methods to eliminate computational errors.

Second, the extracted striped data of the ROI is merged in chronological order to form a new data composite image. To meet the intuitive perspective of data processing, the strip area was rotated and filled with a vertical filling method corresponding to the time axis (horizontal axis) scale. The composite image formed after filling has obvious color layering characteristics, which can visually distinguish the changes in the edge of the upper sample.

Furthermore, in order to highlight the edge features of the image, this chapter uses the threshold segmentation method to binarize the image. The original data form a grayscale image, with each pixel value in the range of [0, 254]. The higher the value, the brighter the pixel. In the process of threshold segmentation,

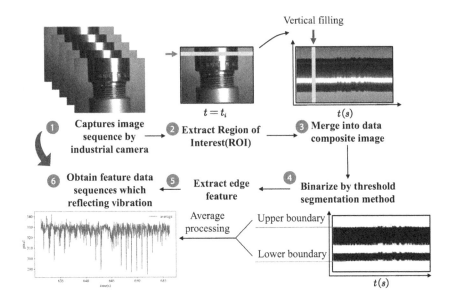

FIGURE 4.93   Image data feature extraction flow.

in order to avoid interference caused by occasional factors such as brightness changes and position changes that may exist during the experimental process, a dynamic threshold method was adopted. The threshold was set to 30% of the overall brightness, with a value of 255 for the parts above the threshold and 0 for the parts below the threshold, achieving the effect of binarization. After threshold segmentation and image binarization, the original boundary attributes have been significantly enhanced, making feature recognition more convenient.

Finally, edge features are extracted from binary images and converted into feature data that reflect the vibration of the upper sample. There are various methods for recognizing edge features. This chapter utilizes the binary image feature to search for boundary features by changing the color (numerical value) of edge features. When the color value changes, the upper and lower boundaries of the upper sample edge are determined by combining the image. Therefore, the obtained data are the pixel position where the upper and lower boundaries are located, and their unit is pixels. Due to the fact that the upper and lower boundary data are the same reflection of the radial vibration of the upper sample, practical feature data sequences are formed by calculating average values and other methods.

4.4.3.3.2   Image data feature analysis

The vibration data extracted from image information come from the radial oscillation of the upper sample fixture, which characterizes the trend of the radial oscillation of the upper sample fixture. The vibration direction is consistent with the direction of friction force. In physical terms, when the characteristic data

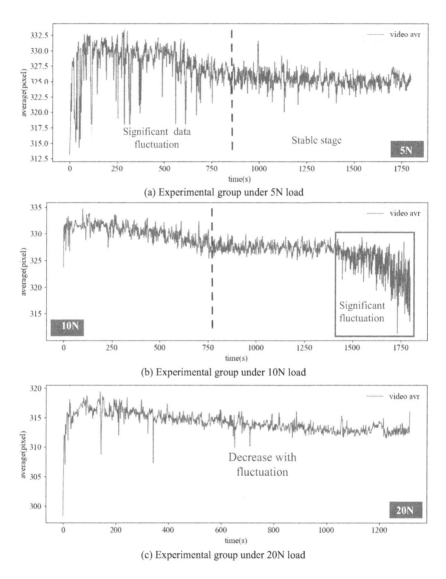

FIGURE 4.94   Change curve of characteristic data reflecting vibration in friction image information.

value is high, it indicates that the overall swing of the upper sample is left when it vibrates radially, and when the value is low, it indicates that the overall swing is right. The vibration information obtained through feature extraction method on the friction test image data obtained by the experimental group over time is shown in Figure 4.94.

From the figure, it can be seen that under different load conditions, the overall vibration data show a trend of first increasing and then decreasing. The physical significance of the corresponding data indicates that the upper sample fixture swings relatively to the left during the initial stage of the friction and wear experiment, and gradually shifts to the right.

By comparing the vibration information (Figure 4.94) with the friction coefficient variation curve (Figure 4.94), it was found that the variation trend of the vibration information extracted from the image signal is highly similar to the variation trend of the friction coefficient, especially under low load conditions, the vibration data fluctuate violently in the early stage and gradually stabilize in the later stage, corresponding to the corresponding friction coefficient variation characteristics one by one. However, under medium load conditions, there was a significant decrease in the later stage of the test process, accompanied by severe fluctuations, which may be related to the random vibration during the friction and wear test process.

In summary, the data results indicate that using the lateral image information of the upper sample fixture collected through experiments, combined with scientific research methods such as deep learning-based motion amplification algorithm, and through image processing, target data that effectively reflect the relationship between friction and wear characteristics can be extracted from the friction image information. This provides a feasible method for promoting friction image research and participating in friction and wear problem research under the framework of friction informatics.

4.3.3.4 Data feature fusion of multi-dimensional friction information

The multi-dimensional friction information obtained from the real-time data collection platform for friction tests has different sampling frequencies and dimensions, which brings inconvenience to the joint analysis and research of multi-dimensional data. By using methods such as time slicing and data dimensionality reduction, data alignment of multi-dimensional friction information can be achieved, which helps to effectively integrate multi-dimensional friction information and facilitates the comparison of data changes without friction information. It is an important prerequisite for utilizing multi-dimensional friction information to achieve correlation analysis, fitting, and prediction of friction coefficients.

In response to the problem of different dimensions of multi-dimensional friction information, this chapter has extracted data features reflecting the radial vibration of the upper specimen from the friction and wear test images through motion amplification algorithms and image processing methods, achieving data dimensionality reduction of two-dimensional images and obtaining a one-dimensional data sequence that changes over time.

However, due to differences in sampling equipment and their sampling principles, there are significant differences in the sampling frequency of multi-dimensional friction information. Among them, the sampling frequency of loading force and friction coefficient is 100 Hz, and the sampling frequency of sound pressure sensor and laser displacement sensor is 1 kHz. The sampling frequency of the digital microphone is 48 kHz, covering the common sound frequency range. The sampling rate of image data is 32 frames per second, which meets the image shooting requirements for the vibration phenomenon of the upper sample. To solve the problem of sampling frequency differences, this chapter adopts a cross-sectional method to process data features, achieving data alignment and effective fusion of multi-dimensional friction information.

The cross-sectional method originated from the study of time series datasets. A time series dataset is a dataset composed of multiple sets of time series data, generally divided into panel datasets and cross-sectional datasets. The panel dataset is composed of multiple sets of time series data that change synchronously over time, and each data group has a cross-section characteristic in the time dimension. When the data have aligned time sections and data are extracted from each time section, a cross-sectional time series dataset can be formed. Panel datasets and cross-sectional datasets have been widely used in the research of time series data in fields such as economics.

During the friction and wear test process, the raw data of the friction and wear test, which is sampled in real time, continuously changes with time and has obvious time series data characteristics. At the same time, the type of data collection is mostly digital signals, and the collection process has discrete characteristics. In terms of time dimension, the experimental data are not continuous and have cross-sectional characteristics with sampling time intervals, thus forming a friction information panel dataset. However, due to the different sampling frequencies of each raw data, the cross-section of the panel dataset is not uniform, which is not conducive to subsequent data analysis.

To overcome the problem of uneven data cross-section caused by the sampling frequency of each sensor, statistical processing such as downsampling and averaging were performed on all feature data at second intervals during data processing such as feature extraction. While obtaining effective features, data alignment in time dimension was achieved, forming a multi-dimensional friction information dataset with a minimum alignment unit of seconds. This dataset achieves effective correspondence between multi-dimensional features and target values (friction coefficients) on a time section, especially through feature extraction and sectionalization, achieving effective fusion of 2D image raw data with other multi-dimensional friction information.

As shown in Figure 4.95, this chapter uses the time slicing method to convert the panel dataset composed of multi-dimensional friction information into a cross-sectional dataset. Then, a data regression model is trained in the time section to effectively fit the target friction coefficient by learning the correlation characteristics between multi-dimensional friction information in the time section and applied to real-time fitting and monitoring of the friction coefficient time series.

From the introduction, it can be seen that there is an inherent correlation between friction information, that is, each piece of friction information has a corresponding relationship in the time dimension. The multi-dimensional friction

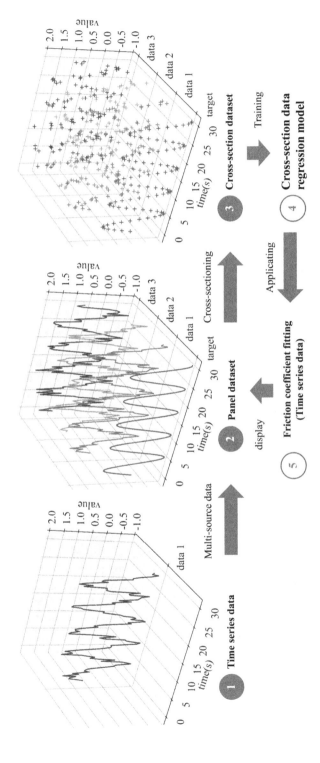

FIGURE 4.95   Tribo-information dataset and cross-section.

information dataset of time section after data fusion provides powerful conditions for effectively utilizing the characteristics of time section data and mining the internal correlation of friction information.

4.3.3.5 Summary

This section utilizes various data processing methods to extract data features from multi-dimensional friction and wear information during the dry friction process. Adopting data filtering and time-frequency domain feature extraction methods, combined with data characteristics and tribological changes, effective feature extraction of various one-dimensional friction information raw data has been achieved, highlighting the trend of friction information changes. Using a learning-based motion amplification algorithm, the radial micro vibration phenomenon of the upper sample was demonstrated, and feature data reflecting vibration information were extracted, achieving data dimensionality reduction of two-dimensional friction image information. Based on the theoretical background of friction phenomena, the changing trends of various friction information were explored, and the effective correlation between vibration signals obtained by image methods and the changing trends of friction coefficients was verified. The effective fusion of multi-dimensional friction information was achieved through the time slicing method, forming a multi-dimensional friction information cross-section dataset, providing a data foundation for subsequent analysis and research.

4.3.4 INTEGRATED MODEL FOR MULTI-DIMENSIONAL INFORMATION FITTING OF FRICTION COEFFICIENT

4.3.4.1 Introduction

Real-time monitoring and prediction of friction coefficient is one of the important objectives of monitoring friction and wear status. The preliminary analysis of the characteristics of friction and wear data shows that there is correlation between friction information and friction coefficient, which is a breakthrough point for real-time monitoring of friction coefficient using friction information. However, the problem is solved by the large number of friction information data samples and complex correlation laws. For the cross-sectional dataset of friction information, this chapter clarifies the linear correlation characteristics between the characteristic data of friction and wear tests through correlation analysis. The regression fitting problem of time section is organically combined with the monitoring problem of time series, and a multiple regression integration model and range evaluation method are established to achieve effective fitting of multi-dimensional friction information to friction coefficient. The effectiveness and accuracy of the fitting model were tested using experimental datasets, providing new ideas for data monitoring of friction coefficients.

4.3.4.2 Correlation analysis of friction information

Feature selection and feature extraction effectively extract data features with stronger correlation and more obvious change trends from the original signal. To further improve the accuracy and reliability of data analysis, this section verifies the correlation characteristics between friction and wear process information in the friction information section dataset through correlation analysis method, based on the preliminary friction and wear test dataset obtained through data processing, providing a data foundation for subsequent regression analysis.

4.3.4.2.1 Dataset correlation analysis

Data correlation analysis is the foundation of regression analysis, and it can also conduct preliminary tests on the correlation characteristics between friction information. For the time aligned friction information cross-section dataset, this chapter conducts correlation analysis using the correlation coefficient matrix method and the data distribution fitting method. The former calculates the correlation coefficients between different friction information in the form of time series, achieving correlation analysis of the overall multi-dimensional friction information data. The latter combines the friction information of the time section into data pairs and performs least squares fitting, realizing the correlation analysis of multi-dimensional friction information in the time section.

Correlation analysis is a universal method for multivariate data analysis. In order to maintain generality, this chapter first abstracts the friction information cross-sectional dataset into a multi row and multi column data matrix, thus conducting a detailed discussion of correlation analysis methods. Assuming that the feature data has m rows and n columns, a data matrix can be formed as shown in Equation 4.85.

$$X = \begin{bmatrix} x_{11} & x_{12} & \cdots & x_{1n} \\ x_{21} & x_{22} & \cdots & x_{2n} \\ \vdots & \vdots & \ddots & \vdots \\ x_{m1} & x_{m2} & \cdots & x_{mn} \end{bmatrix} \tag{4.85}$$

In the equation, each row of features represents a sample, which corresponds to a unit time in the dataset of this chapter. Regarding the correlation between columns a (X_a) and b (X_b) in the data matrix, the correlation coefficient is set to $r(a, b)$, and the Pearson correlation coefficient is calculated according to Equation 4.86.

$$r(a, b) = \frac{\sum_{i=1}^{m}(x_{i,a} - \overline{x_a})(x_{i,b} - \overline{x_b})}{\sum_{i=1}^{m}(x_{i,a} - \overline{x_a})^2 \sum_{j=1}^{m}(x_{j,b} - \overline{x_b})^2} \tag{4.86}$$

After traversing all data columns in Equation 4.86, a and b can obtain the correlation coefficients between each column in the dataset, forming a correlation coefficient matrix. In the matrix, the values of each correlation coefficient

$r(a, b)$ are within the range of $[-1,1]$. The closer the absolute value is to 1, the stronger the linear correlation between the data. If the correlation coefficient is positive, it indicates that the two sets of data are positively correlated, and if it is negative, it indicates that the data are negatively correlated.

Using the Pearson correlation coefficient calculation equation, calculate the correlation coefficients between each piece of data in the friction information dataset. Due to the similarity in data characteristics between laser displacement sensors and industrial cameras, with the same research objectives and redundant information, they form multi-dimensional friction information datasets with other data, calculate correlation coefficients, and participate in subsequent data processing and analysis.

4.3.4.2.1.1 Correlation coefficient matrix of multi-dimensional friction information feature data   For a multi-dimensional friction information dataset without image information, the correlation coefficient matrix is shown in Figure 4.96. The values in the matrix represent the correlation coefficients between two datasets on the horizontal and vertical axes, and can also be visually displayed through color. Dark colors indicate positive correlation, while light colors indicate negative correlation.

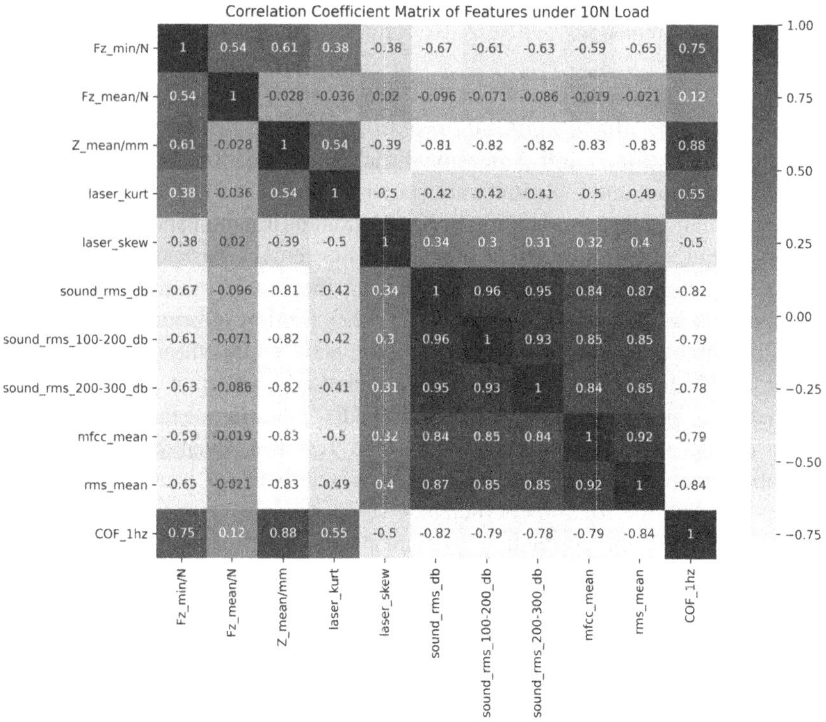

FIGURE 4.96   Correlation coefficient matrix of feature data.

The results indicate that there is a high linear correlation between the friction and wear test data. In the experimental data, there is a strong positive correlation between the target friction coefficient (No. 11) and the loading force (No. 1 and 2), indicating that when the friction coefficient is large, the characteristics of the loading force data reflect an intensification of the vibration of the upper sample fixture. The height value of the upper sample (No. 3) reflects the wear depth. As the test time increases, the wear depth increases, while the height of the upper sample decreases. Therefore, there is a positive correlation between the value and the friction coefficient. The friction coefficient is positively correlated with the kurtosis (No. 4) of the vibration curve of the upper sample, and negatively correlated with the skewness (No. 5), indicating that the vibration signal is sharp in the early stage of the friction and wear process, and the vibration phenomenon is obvious. As the friction and wear gradually evolve, the vibration phenomenon gradually tends to be gentle.

At the same time, there is a significant negative correlation between the friction coefficient and all sound signal features, which may be related to the stage and state changes of friction wear. As wear intensifies, the friction noise increases, and the friction coefficient decreases as it enters a stable friction stage. In addition, there is a strong correlation between multiple features related to sound, indicating that the changing characteristics of sound information correspond to each other, demonstrating consistency in sound change trends from different dimensions.

4.3.4.2.1.2 Multi-dimensional friction information correlation coefficient matrix containing image information feature data   Calculate the correlation coefficient matrix for a multi-dimensional friction information dataset containing image information, and the results are shown in Figure 4.97.

The correlation characteristics between friction information and loading force, friction sound, and other friction information remain basically unchanged. Observing the correlation coefficient between the image information (No. 4) of the motion amplification algorithm and other friction information, it can be found that there is a strong positive correlation between this information and the friction coefficient (No. 10), with a correlation coefficient of 0.79. This information is another representation of the radial vibration of the upper sample fixture during the friction process, and the result is even better than the correlation coefficient (0.55) between the characteristics of laser displacement sensing data and the friction coefficient.

The analysis of the correlation coefficient matrix of feature data containing image information shows that effective extraction of data information with strong correlation with target values such as friction coefficient in image information has been achieved through scientific and technological methods, which helps to promote the participation of friction information such as images in the study of friction and wear processes and helps to analyze and predict related data using methods such as linear regression.

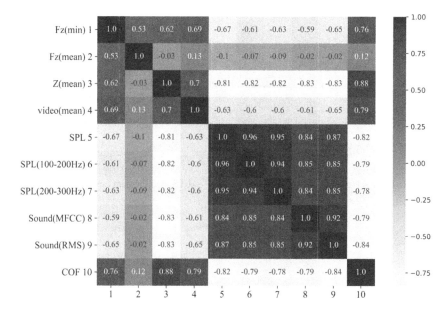

FIGURE 4.97 Correlation coefficient matrix of feature data with image information.

4.3.4.2.2 Data feature distribution fitting

After cross-sectional analysis of the friction information dataset, two dimensions of correlation characteristics research problems were formed: the correlation between the changing trends of each piece of time series data in the time dimension, and the correlation between each piece of friction information data in the time section. The correlation coefficient matrix method starts from the overall data and calculates the correlation between friction information time series in the time dimension, solving the first research problem. The distribution fitting method of data features studied the correlation between various friction information data in the time section to solve the second problem.

If the influence of time factors on data changes is ignored, the friction information and target value are corresponded one by one to form a data combination, and represented in the form of a scatter plot to obtain the data feature distribution between the friction information and the target value, as shown in the gray data points in Figure 4.98. The horizontal axis in the figure represents the friction coefficient (fitting value), and the vertical axis represents the monitoring value of friction information. By using methods such as least squares to fit the data distribution, it can more intuitively reflect the relevant characteristics between friction information, and thus test the corresponding relationship of data features in the time section.

4.3.4.2.2.1 Least squares method
Least squares is an optimized data fitting method that minimizes the sum of squares of errors to find the optimal fitting function for existing data.

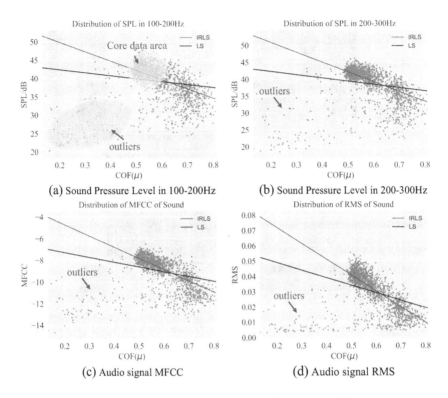

FIGURE 4.98   Distribution fitting of cross-sectional feature data (5N).

The data samples are shown in Equation 4.10, and the target value data corresponding to each sample forms a label matrix, as shown in Equation 4.87.

$$Y = [y_1 \quad y_2 \quad \cdots \quad y_m]^T \tag{4.87}$$

The goal of the least squares method is to find the optimal set of parameters θ, as shown in Equation 4.88.

$$\theta = [\theta_1 \quad \theta_2 \quad \cdots \quad \theta_n]^T \tag{4.88}$$

thus, establishing a fitting relationship between the sample and the label as shown in Equation 4.89.

$$X\theta \sim Y \tag{4.89}$$

At this point, the sum of squares of the fitting results is represented by Equation 4.90.

$$f(\theta) = \frac{1}{2}\|X\theta - Y\|^2 \tag{4.90}$$

The least squares method achieves parameters by minimizing the sum of squares of the above errors θ. The optimal analytical solution can be obtained by solving the partial derivative method, as shown in Equation 4.91.

$$\theta = (X^T X)^{-1} X^T Y \tag{4.91}$$

In the case of a small amount of data, the analytical solution shown in Equation 4.91 can be obtained through matrix operation, and each sample error has an equal contribution to parameter optimization. However, the friction information dataset generated by each formal experiment can form a data matrix of approximately 1800 rows and 10 columns after sectioning and contains outliers caused by randomness and other factors. If each sample still maintains the same contribution in the fitting model, these abnormal data will obviously interfere with the fitting model, leading to significant errors in the fitting results. As shown in Figure 4.98, the blue fitting line is the distribution fitting result achieved by the least squares method. The fitting line is clearly disturbed by sparse outliers in the lower left part of the figure and does not accurately depict the trend of changes in the main data.

4.3.4.2.2.2 Weighted least squares method To effectively overcome issues such as outlier interference, the weighted least squares (WLS) method is based on the conventional least squares method by introducing a weight parameter matrix as shown in Equation 4.92:

$$W = \begin{bmatrix} w_1 & 0 & \cdots & 0 \\ 0 & w_2 & \cdots & 0 \\ \vdots & \vdots & \ddots & \vdots \\ 0 & 0 & \cdots & w_m \end{bmatrix} \tag{4.92}$$

By adjusting the contribution of each sample error in the parameter optimization process, the stability of the fitting model is enhanced, and it is a robust linear fitting model (RLM). At this point, the weighted sum of squares of the fitting results becomes as shown in Equation 4.93.

$$f(\theta) = \frac{1}{2}\|W(X\theta - Y)\|^2 \tag{4.93}$$

However, the weight parameter matrix of this method relies on manual selection, and when the data sample size is large, the selection of weight parameters faces problems such as objectivity and significant reduction in efficiency.

4.3.4.2.2.3 Iterative reweighted least squares method To solve the problem of selecting weight parameters under large sample data, on the basis of weighted least squares method, iterative reweighted least squares (IRLS) continuously optimizes the sample weight parameters, thereby gradually optimizing the error to the given robust error estimation value.

The core of the IRLS method is to update the weights through Equation 4.94:

$$w_i^{(s)} = \frac{1}{|y_i - X_i\theta^{(s)}|} \tag{4.94}$$

In the equation, s represents the current iteration step, y_i represents the i-th label, x_i represents the i-th sample.

Before iteration, the initial value of the weight w_i is usually set to 0. During the iteration process, the fitting error is continuously utilized to correct the sample weight, ultimately achieving robust linear model estimation. To avoid dividing by 0, a threshold will be set in the actual process δ (usually 0.0001), compared with the error value to form a practical weight update method, as shown in Equation 4.95.

$$w_i^{(s)} = \frac{1}{\max\{\delta, |y_i - X_i\theta^{(s)}|\}} \tag{4.95}$$

4.3.4.2.2.4 Data distribution fitting After discussing the above methods, in order to effectively address issues such as abnormal values of friction information dataset samples under the influence of randomness in friction and wear tests, and achieve more stable feature distribution fitting of friction information data, the IRLS method should be selected for feature distribution fitting of cross-sectional data. This chapter focuses on the correlation between frictional acoustic information and friction coefficient in the time section. Therefore, data distribution fitting research is conducted for various frictional acoustic information and friction coefficients.

In the process of data distribution fitting, different friction information data are used as samples, and target values (friction coefficients) are labeled to form multiple sets of data pairs to be fitted, forming a scatter data basis for friction coefficients (horizontal axis) and various friction information feature data (vertical axis). Then, the IRLS method is used for data fitting to obtain linear distribution fitting results. Figures 4.98–4.100 show the fitting of data distribution for different load test groups.

The results indicate that under different load conditions, there is a significant linear negative correlation between the friction coefficient and the characteristic values of each sound information.

In terms of details, under low load conditions (5N), there are many outliers, and the IRLS method is more effective in fitting the main data than the least

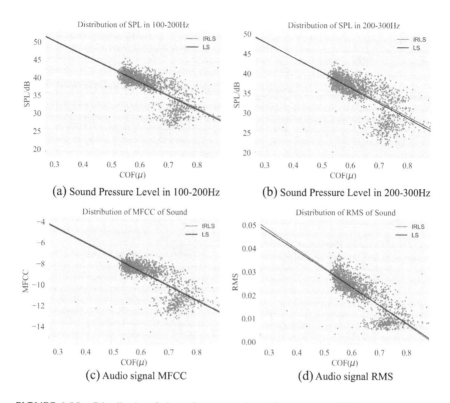

(a) Sound Pressure Level in 100-200Hz   (b) Sound Pressure Level in 200-300Hz

(c) Audio signal MFCC   (d) Audio signal RMS

FIGURE 4.99   Distribution fitting of cross-sectional feature data (10N).

squares method. The fitting lines of the two methods have significant differences. In contrast, under medium to high load conditions, as shown in Figures 4.99 and 4.100, outliers are reduced, and data are more concentrated. The fitting results of IRLS method and least squares method are relatively close.

In addition, it can be seen from the distribution fitting diagram that under different load conditions, the range of 0.5 to 0.6 of the friction coefficient (horizontal axis) is the area with the most concentrated data points, namely the core data area. Under low load conditions (5N), the core data area data points account for about 80% of the total data, because the sliding friction coefficient of metal parts has certain regularity and stability from a statistical perspective. However, due to the continuous changes in wear status during the friction process, there are also significant fluctuations in the friction coefficient test values, resulting in discrete data points outside the core data area. As the load increases, the friction contact pair fits more tightly and the fluctuation decreases. At this time, the friction coefficient and SPL still exhibit a linear distribution feature, effectively demonstrating the correlation between friction information and facilitating the use of friction information data for linear regression fitting of friction coefficient and other data.

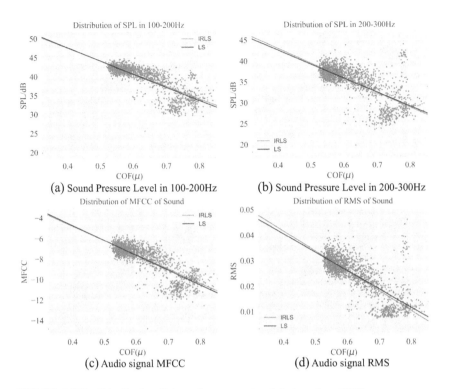

(a) Sound Pressure Level in 100-200Hz

(b) Sound Pressure Level in 200-300Hz

(c) Audio signal MFCC

(d) Audio signal RMS

FIGURE 4.100   Distribution fitting of cross-sectional feature data (20N).

In summary, this case preliminarily verifies the data correlation between friction information and friction coefficient by calculating the Pearson correlation coefficient matrix between friction information, indicating the excellent linear correlation features of the friction information dataset. Subsequently, the linear correlation between friction noise and friction coefficient was fitted using an iterative weighted least squares fitting method, which strongly demonstrated the significant negative correlation between the two. At the same time, the discrete point characteristics of friction and wear test data were observed, which will be further addressed through data processing and analysis.

4.3.4.3   Fitting regression analysis method

Regression analysis is a mathematical method of modeling and analyzing the correlation between variables, which can be used to predict variables using regression models. This chapter collects friction information data while also collecting the target value friction coefficient, with the aim of evaluating and verifying the prediction results of the regression model. By accurately fitting and estimating the known target values, an effective prediction model is established, which is the center of this chapter's exploration of friction coefficient monitoring methods.

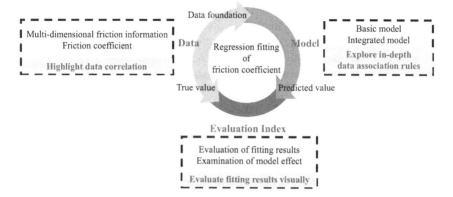

FIGURE 4.101   Regression fitting process of friction coefficient.

A complete friction coefficient fitting regression model consists of three parts: data, model, and evaluation indicators. Data provide the foundation for the model, and evaluation indicators test the model. The three work together to form the complete process of friction coefficient fitting regression, as shown in Figure 4.101.

In statistical research, the correlation of data is the foundation of linear regression analysis. This chapter extracted friction information with obvious phenomena and significant changes during the experimental process and enhanced the correlation of the data through feature selection and feature extraction. The high correlation between friction information feature data and friction coefficient has been fully verified in the correlation analysis, providing a solid data foundation for the regression analysis of friction information to friction coefficient.

The friction and wear cross-section dataset has a large sample size, multiple types of data, and complex correlation rules. At the same time, the friction and wear test data have a certain degree of randomness. In order to deeply explore the correlation characteristics between multi-dimensional friction information and friction coefficients. On the one hand, this case starts from various basic regression models and gradually establishes complex regression models with universal applicability and accurate and reliable fitting results. On the other hand, this case defines evaluation indicators suitable for friction coefficient regression analysis to objectively evaluate the fitting effect.

4.3.4.3.1 Evaluation indicators

The fitting effect of regression models on the test set is often evaluated using indicators such as mean square error (MSE), root mean square error (RMSE), and goodness of fit (R^2). Let the fitting value be $\hat{Y}$. The true value is Y, each containing m numerical values, as shown in Equation 4.10. Therefore, the existing evaluation index equation is as follows:

1. The mean square error focuses on evaluating the absolute error between the fitted value and the true value. Its calculation method is shown in Equation 4.96, and the result is represented as the average of all absolute errors. A smaller value indicates a smaller error.

$$MSE = \frac{1}{m} \Sigma_{i=1}^{m} (y_i - \hat{y}_i)^2 \qquad (4.96)$$

2. The root mean square deviation was calculated based on the mean square deviation, as shown in Equation 4.97, achieving consistency between the evaluation indicator dimension and the data dimension. The smaller the value, the smaller the error;

$$RMSE = \sqrt{\frac{1}{m} \Sigma_{i=1}^{m} (y_i - \hat{y}_i)^2} \qquad (4.97)$$

3. The goodness of fit (R^2) reflects the degree of interpretation of the fitting relationship, and the calculation method is shown in Equation 4.98. The closer the value is to 1, the better the fitting effect of the model.

$$R^2(y, \hat{y}) = 1 - \frac{\Sigma_{i=0}^{m} (y_i - \hat{y}_i)^2}{\Sigma_{i=0}^{m} (y_i - \bar{Y})^2} \qquad (4.98)$$

The potential premise of the above evaluation indicators is that the fitted values should be completely equal to the true values, and the accumulation of errors caused by outliers will significantly interfere with the evaluation indicators. However, there is inevitably a certain degree of randomness in friction and wear tests. Under the same conditions, the data characteristics exhibited by test data such as friction coefficient are that there is a reasonable interval rather than a fixed value. Therefore, when using experimental data as a reference value, a reliable range should be used instead of a single value to avoid fitting errors caused by the randomness of experimental data. At the same time, it is necessary to avoid excessive degradation of the fitting effect caused by outliers and reduce the impact on the average error.

In response to the above issues, an indicator for evaluating the fitting effect of regression models, coverage ratio (CR) of fitting values, is proposed. Compared to the evaluation indicators of linear regression problems, CR has the characteristics of adaptive range evaluation and independent evaluation, meeting the evaluation requirements for fitting accuracy. On the other hand, through fuzzification methods, it pays better attention to the consistency between the fitted values and the true values, making it suitable for real-time monitoring of friction coefficients. According to the fitting accuracy and target value, the error interval is obtained. When the fitting value is within the error interval near the target value, it is considered that the fitting value meets the fitting accuracy. The proportion of all fitting values meeting the fitting accuracy is the CR of the fitting value.

The specific calculation equation for fitting numerical CR is shown in Equation 4.99 to Equation 4.101.

$$e_i = (1 - a)y_i, \ i = 1, 2, \cdots, m \tag{4.99}$$

$$\delta_i = \begin{cases} 1, f_i \in [y_i - e_i, y_i + e_i] \\ 0, others \end{cases}, \ i = 1, 2, \cdots, m \tag{4.100}$$

$$CR = \frac{\sum_{i=1}^{m} \delta_i}{m}, \ i = 1, 2, \cdots, m \tag{4.101}$$

a is the fitting accuracy.

y_i is the target value of the i-th experimental data.

e_i is the maximum allowable error of the i-th value.

f_i is the fitting value of the i-th experimental data target.

δ_i represents whether the fitting error of the i-th experimental data target meets the maximum allowable error.

m is the total number of target value samples for the experimental data.

CR is the coverage ratio of fitted values.

The use of the fitting result CR indicator can effectively enhance the model's tolerance for abnormal data and randomness during fitting result analysis, which helps the model accurately identify the main features of the data. As shown in Figure 4.102, data points containing random errors are used as fitting values, and

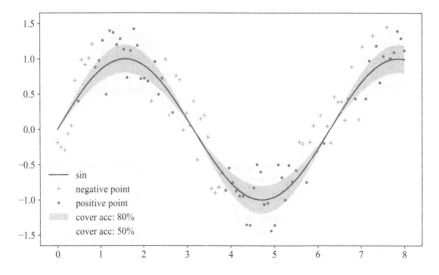

FIGURE 4.102   The effect of the coverage ratio of the fitting result.

sine curves (solid lines in the figure) are used as actual values. Based on the fitting accuracy, allowable error intervals (shaded areas in the figure) are set to calculate the CR evaluation index value. The CR evaluation index will calculate the number of fitting points within the allowable error range and divide it by the total number of all fitting points to infer the similarity between the current data sample and the sine curve.

From the perspective of the image, the CR evaluation indicator shows the coverage of the allowable error interval on the fitting points, so this indicator is named the fitting result CR. The CR evaluation index avoids excessive interference of outliers on data fitting and effectively enhances the stability of the evaluation method with a range evaluation strategy. It is very suitable for evaluating the fitting results of friction and wear test data with certain randomness and error interference.

This chapter sets the fitting accuracy to 95%. According to the calculation Equation 4.99 to Equation 4.101, 5% of the experimental value is taken as the maximum allowable error, and the fitting value CR index is calculated for the regression model.

4.3.4.3.2 Basic model

In the long-term research of data fitting problems, numerous basic regression models with different concerns have been continuously developed, which are enlightening and guiding in dealing with complex regression problems. Especially when dealing with the multicollinearity problem within the dataset, which has correlations, the fitting effect of conventional linear regression methods will be greatly reduced. Many existing basic models have adopted different methods to overcome this problem.

This chapter has used six basic regression models for preliminary regression of data, namely ridge regression method (Ridge) based on linear regression method, eastic net (EN) regression method, KRR method, and Bayesian Ridge Regression (BRR) method, SVR based on SVM method and extreme random tree (Extra) regression method based on the RF method can be used for multivariate regression problems.

Compared to the linear regression (LR) method, the ridge regression method (Ridge) increases the L2 regularization penalty term as shown in Equation 4.27. Penalize the behavior of overemphasizing the role of a single sample to reduce fitting errors, effectively improving multicollinearity and overfitting problems.

$$\varphi_2 = \sum_{i=1}^{m} |w_i^2| \qquad (4.102)$$

The elastic network (EN) regression method combines L1 and L2 regularization penalty terms, taking into account the advantages of lasso regression and ridge regression, by using the L1 regularization term φ_1 as shown in Equation 4.28. It selects more valuable predictive variables while preserving the adaptability of ridge regression to multicollinearity data.

$$\varphi_1 = \Sigma_{i=1}^m |w_i| \qquad\qquad (4.103)$$

The KRR method, based on the ridge regression method, utilizes kernel function methods to enhance the nonlinear characteristics of training samples. It is a multivariate nonlinear regression analysis method and has advantages in nonlinear fitting problems.

The BRR method considers that the parameters of the fitted model also satisfy the Gaussian distribution, and calculates the posterior through the prior of the model parameters, enhancing the stability of the regression model.

Compared to linear regression methods, SVR is based on SVM methods rather than least squares methods and does not need to worry about the impact of multicollinearity. It pays more attention to the impact of support vectors on fitting the model and has better fitting effects on the banded area.

The Extra method is based on decision trees. Compared to the RF method, it has stronger randomness when selecting eigenvalues to partition the decision tree. Although the generated decision tree size will be larger, the model's generalization ability has also been further improved, and it can handle high-dimensional feature data well.

The adaptability of different models to data patterns varies. On the one hand, it is necessary to combine the characteristics of friction information data and the fitting ability of each model to the friction and wear experimental dataset for reasonable selection. On the other hand, the method of integrating multiple models can be used to establish integrated regression models based on multiple models, in order to achieve the best data regression fitting effect.

4.3.4.3.3 Model selection

Due to the different characteristics of the models, the fitting effects of different models are uneven, making it difficult to achieve good fitting accuracy and also hindering the establishment of a multi method integrated friction information correlation law regression model. Therefore, the selection method should be used to screen from multiple models.

In the study of regression fitting problems, a simple multi method fusion technique is to weight the fitting results of multiple models, take the average, and obtain new fitting results. From a statistical perspective, the weight average model helps to utilize the fitting results of excellent models to reconcile the fitting errors of poor models and enhances the stability of the results through various regression models. It is a fast and convenient complex regression model. However, the weighted regression model has not significantly improved its ability to analyze data, nor has it achieved in-depth exploration of data association rules. It may even drag down excellent models due to data errors in poor models, resulting in a phenomenon where the fitting results are not as good as the basic model after weighted average.

Although weighted regression models are not an ideal method for studying complex data patterns, this phenomenon provides ideas for the optimization of multiple models. In a weighted regression model, the weights of the fitting results of each basic model can be generated in various ways. This chapter uses evaluation values for the fitting results of the basic model, assigns weights to the fitting data of each model, and calculates the weighted fitting average. The optimization of the model can be achieved by directly comparing the weight values or comparing the fitting effects of the basic model and the weighted regression model.

4.3.4.4   Integrated regression model for friction coefficient

The friction information dataset has the characteristics of multi-dimensional and large sample size, and traditional multiple linear regression methods are difficult to learn its complex features. Optimizing the regression model from the perspective of dataset utilization methods and integration of multiple basic models can achieve better fitting results.

4.3.4.4.1   Integrated regression model

The optimization of fitting regression models also requires consideration of the interaction between data and models. Through collaborative optimization of data and models, the effect of integration is greater than summation.

To ensure the reliability of fitting regression models, the set aside method is usually used to divide the dataset into training and testing sets. The testing set will not participate in the training process and will only be used to validate the trained model. The model will be fully trained on the basis of the training set, learning the data rules of the training set, and tested on the test set. The hidden problem is that the training set often has a large scale and directly uses all data for model learning, making it difficult to fully understand the regular features in the data samples and easily ignoring the deep correlation rules of the data.

Therefore, in order to make more full use of the dataset, the K-fold cross-validation method was used to further divide the training set into K segments. During the model training process, one segment was used in turn as the test set, and the remaining segments were used as the training set. During the training process, the testing of learning effectiveness was strengthened, and the mining of association rules in the dataset was more detailed, improving the generalization ability of the regression model.

Furthermore, based on the K-fold cross-validation method of the dataset, a stacking ensemble learning method is adopted to further explore the features of friction information data. This method is divided into two layers of models. In the first layer, multiple basic models are used separately, combined with the K-fold cross-validation method for learning. The predicted results of each model will be merged into the input data of the second layer model. In the second layer, a new basic model is used for relearning, and deep correlation rules are extracted from the prediction results of the previous layer.

When applied to the regression problem of fitting friction coefficients with multi-dimensional friction information, friction information is used as the input value of the first layer model. By learning multiple basic regression models,

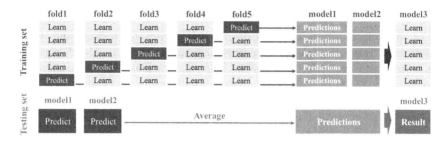

FIGURE 4.103   K-fold cross-validation two-layer stacking model.

multiple sets of preliminary fitting values of COF can be obtained. Using the initial fitting values of multiple models as the input values for the second layer model, learn the new regression model again to obtain accurate fitting results for COF. In the resulting two-layer model, the first layer focuses on the contribution of different friction information to COF fitting using different methods, while the second layer focuses on the accuracy of different models in COF fitting. Compared with the single-layer linear regression model, it can obtain data features more deeply, which is helpful for regression learning of multi-dimensional large sample data. The integrated double-layer stacked regression model is shown in Figure 4.103.

The K-fold cross-validation and double-layer stacking methods provide a way to integrate multiple basic models.

Using various machine learning methods such as Ridge, EN, KRR, BRR, SVR, Extra, etc., the data are preliminarily fitted without the use of K-fold cross-validation. By adjusting hyperparameters, the fitting effect of each model is optimized. Then, a weighted average model is used as a comparison to select the regression model with better fitting effect.

After statistics, the number of times the model results are superior to the weighted average fitting effect is shown in Table 4.24. Among them, the Extra and BRR methods are significantly superior to other methods and have shown good fitting performance on multiple experimental datasets. Therefore, in the double-layer stacked regression model, the first layer selects the ETR and BRR methods with the best fitting results, and the second layer selects the EN model.

TABLE 4.24

Weighted Average Fitting Model Selection Statistics

Model	Ridge	SVR	KRR	BRR	EN	Extra
>Weight Average*	0	0	0	2	0	5

Notes

* Representing the number of times that the model has better fitting effect than weighted average model, and there are six groups of test data in total.

4.3.4.4.2   Fitting results and analysis

Train the model on a time cross-sectional dataset of friction and wear information features, with the experimental group as the unit. The dataset for each group of experiments was divided using the retention method in the time dimension, with the first 80% of the data being the training set and the last 20% being the testing set. For each training set, the K-fold cross-validation method and double-layer stacked regression model were used to obtain the fitting results of the friction coefficients of each experimental group, and the fitting results were evaluated through the CR index.

As mentioned earlier, the vibration information extracted from the image is redundant and cross-referenced with the vibration information obtained from the laser sensor. Therefore, this chapter focuses on fitting the friction information data containing laser sensor vibration information and the friction information data containing image information and discusses the fitting results of the friction coefficient.

4.3.4.4.2.1 Fitting friction coefficient with multi-dimensional friction information including laser sensor vibration information   The comparison of CR evaluation indicators for friction coefficient fitting results based on friction information data containing laser sensor vibration information obtained from multiple experiments is shown in Table 4.25. When the fitting accuracy is 95%, the fitting data coverage of the Stacking Model can reach over 98% under different load conditions, reflecting the stability of the model in fitting experimental data.

At the same time, through the stacked multiple model integration method, better fitting accuracy was achieved than the various basic models, especially under low load (5*N*) conditions. The fitting accuracy of the stacked model was improved by 17–39% compared to the various basic regression models, reflecting the superiority of the integration method in dealing with complex data features. In addition, the Extra method in the basic model also showed good fitting results,

TABLE 4.25

Coverage Ratio of Fitting Results of Each Model

Coverage Ratio 95%	5 *N*	10 *N*	20 *N*
Ridge	0.5950	0.9412	0.8163
SVR	0.6604	0.8613	0.9541
KernelRidge	0.6667	0.8025	0.9745
BayesianRidge	0.5950	0.9664	0.8622
ElastciNet	0.6106	0.8992	0.8316
ExtraTreeRegressor	0.8131	0.9874	0.9949
Weight Average	0.6791	0.9538	0.9796
Stacking(the chosen model)	0.9844	0.9916	0.9949

indicating its strong generalization ability and good adaptability to friction coefficient fitting problems. Especially under high load (20N) conditions, the fitting evaluation results are the same as those of the integrated model.

The comparison of fitting results under low load is shown in Figure 4.104. The actual measured value curve of COF in the figure is labeled as "origin," and the fitting value curves of each regression model are annotated with abbreviations. Under low load (5N) conditions, the stacked model has higher fitting accuracy compared to the basic model, is closer to real friction information data, has fewer mutations, and effectively reflects the trend of friction coefficient changes.

In response to multiple changes in the peak shape of the friction coefficient (in the box area in Figure 4.104), the error between the stacked model and the experimental values is smaller, and the fitting effect is better, achieving effective tracking of the trend of COF changes. Especially during the rapid rise and fall of multiple friction coefficients with time axis scales of around 75 s, 180 s, and 300 s, compared to the basic model, the fitting error of the integrated model is significantly reduced, indicating that the integrated model has effectively learned the deep correlation of data and achieved better data fitting stability.

As shown in Figure 4.105, under the condition of medium load (10N), the fitting accuracy of the stacking model and the Extra method is close. The reason is that in the second layer of the model, the output value (model fitting value) of the first layer model is used as the new input, and the regression fitting of the friction coefficient is performed again. As the regression fitting reflects the participation of each sample in the fitting results, the double-layer stacked model has a certain model selection ability. The Extra method has demonstrated excellent fitting ability and data generalization ability in the fitting process of each group of experimental data; therefore, it has a prominent guiding role in the integrated model.

In terms of details, in the fast fluctuation range of friction coefficient around 180 s, the stacked model data curve achieves effective tracking of the original data, which is relatively close to the results obtained by the extreme random number method. However, in the interval close to the tail (300–350 s), the integrated model generated fitting errors similar to the basic model, because the stacked model was built on top of the basic model, both basic models did not obtain accurate results, and the two were relatively close, thus failing to provide effective feedback to the deep model. However, the curve can still indicate that the stacked model has responded effectively to similar short-term peak changes.

As shown in Figure 4.106, under high load test conditions, both the basic model and the integrated model exhibit a certain tendency of overfitting. In the 50 to 300 range of the time axis scale, for long-term short and flat fluctuation trends, the fitting accuracy is not high, but the fitting values are basically kept near the experimental data. In contrast, in areas with significant fluctuations in friction coefficient (box areas), such as 0 to 50 s and 300 to 350 s, the model exhibits a stable and positive fitting effect, and the fitting values also show significant changes, especially in the 300 to 350 s range. The fitting effect of the

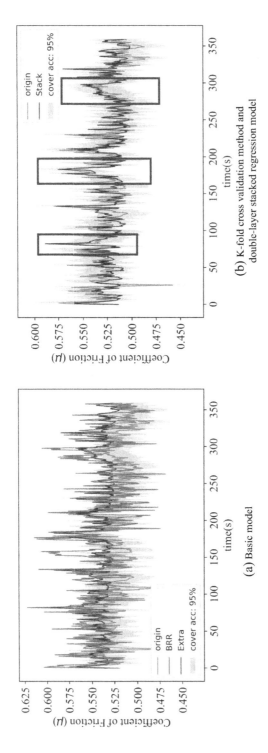

FIGURE 4.104 Comparison of fitting results of low load (5N) test data.

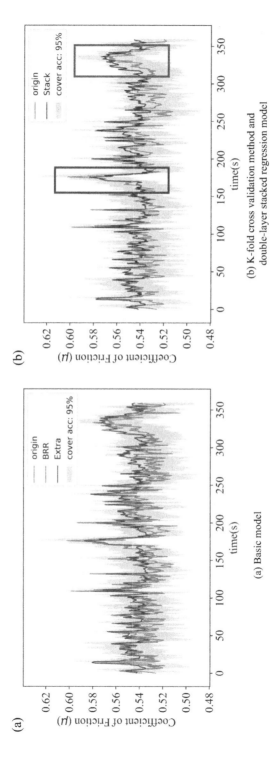

FIGURE 4.105   Comparison of fitting results of middle load (10*N*) test data.

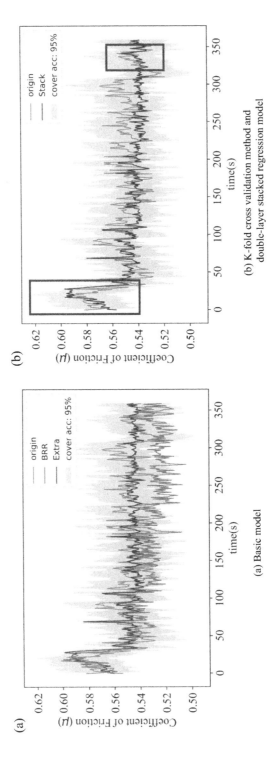

FIGURE 4.106 Comparison of fitting results of high load (20*N*) test data.

stacked model is significantly improved compared to the two basic models, which is closer to the original data, indicating that the basic model has formed effective data feedback to the stacked model.

4.3.4.4.2.2 Multi-dimensional friction information fitting with image information features for friction coefficients

For the image information obtained from multiple experiments, data features reflecting vibration were obtained through motion amplification algorithm. A new friction information dataset was formed by replacing laser displacement sensing data. The friction coefficient fitting results were fitted using this set of data, and the CR evaluation index was calculated. The results are shown in Table 4.26.

Compared to various basic models, the stacked integrated regression model (Stacking) also achieved stable and accurate fitting results. Except for the medium load test data, the fitting effect was significantly improved, and in the low load test group, it also showed excellent fitting effect.

However, there are significant differences in the fitting performance of various basic models under different load conditions. For example, the SVR model performs very well under a $10N$ load, but its fitting performance is greatly reduced under other load conditions, especially the KRR model. When dealing with friction information data containing image information, its fitting performance is relatively poor. The above differences in the fitting effect of the basic regression model indicate that the model and data complement each other. The integrated model achieves stable fitting results under different loads, breaking through the constraints of the model on data, and is a universal regression fitting model.

As shown in Figure 4.107, under low load ($5N$) conditions, after replacing the features of laser displacement sensing data with the feature information in the image signal, the stacked model also has high accuracy in fitting the data, demonstrating the universal applicability of the model to different data features and the stability of the fitting effect.

TABLE 4.26
Coverage Ratio of Fitting Results of Each Model

Coverage Ratio 95%	5 N	10 N	20 N
Ridge	0.4528	0.5556	0.8977
SVR	0.4750	0.9139	0.4924
KernelRidge	0.3694	0.7388	0.2197
BayesianRidge	0.6667	0.4389	0.9205
ElastciNet	0.4639	0.8972	0.7500
ExtraTreeRegressor	0.7778	0.9944	0.9129
Weight Average	0.6083	0.9167	0.9129
Stacking(the chosen model)	0.9194	0.9028	0.9432

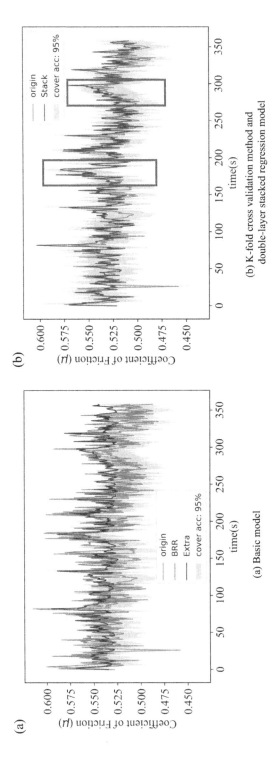

FIGURE 4.107 Comparison of the fitting results of the low load (5N) group with image information.

As shown in Figure 4.108, under medium load (10N) conditions, the stacking model fully absorbs the fitting advantages of two basic models and has smaller fitting errors when the friction coefficient fluctuates significantly, which better reflects the true values and changing trends. The Extra method exhibits obvious overfitting characteristics, focusing on peak changes without effectively fitting valley values. The experimental results fully demonstrate that the integration of multiple models has achieved the research goal of absorbing the strengths of each method and supplementing its weaknesses.

As shown in Figure 4.109, under high load (20N) conditions, the addition of vibration features in the image information significantly improves the data fitting effect. The fitting curve of the friction coefficient is closer to the better fitting effect of the BRR method in the basic model, but in some details, the fitting results are positively adjusted by the Extra method, resulting in better fitting accuracy.

The above fitting results highlight the following characteristics of the stacked model in the process of fitting friction coefficients using friction information:

1. The integrated model is not easily affected by experimental parameters, and its fitting effect is stable under different experimental conditions, with strong universality, especially suitable for the study of friction coefficient fitting under low load conditions.
2. The integrated model has a significant fitting effect on the trend of significant fluctuations in friction coefficient and has excellent tracking ability, which helps to achieve data monitoring of abnormal changes in friction coefficient.
3. The integrated model fully absorbs the fitting advantages of various basic models and effectively adjusts the fitting errors of each basic model, achieving the mining and application of deep correlation features between friction information and friction coefficients.

Finally, the above research results indicate that the fitting model learned from the training set of the stacked model can be transferred to the test set. Due to the chronological relationship between the training and testing sets in the same set of experimental data, this method demonstrates the continuity of friction information correlation features in the time dimension, providing support for predicting friction coefficients and other data using friction information during friction and wear testing.

4.3.4.5 Summary

This section first utilizes a cross-sectional friction information dataset to transform the problem of monitoring the friction coefficient from a time series perspective into a data fitting problem for the time section. Furthermore, the correlation analysis methods such as correlation coefficient matrix and data distribution fitting were used to test the correlation of friction information feature data, providing a theoretical basis for fitting regression problems. Finally, an integrated regression model based on multiple machine learning methods was established to solve the problem of fitting friction coefficients with

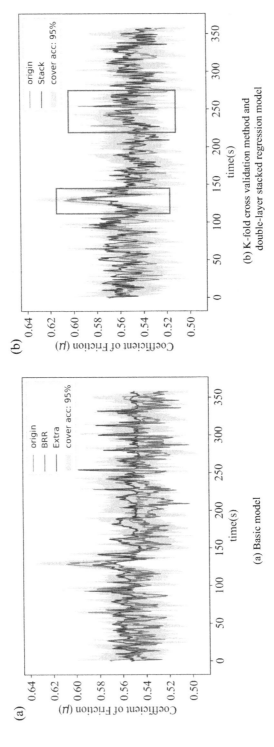

FIGURE 4.108 Comparison of the fitting results of the middle load (10N) group with image information.

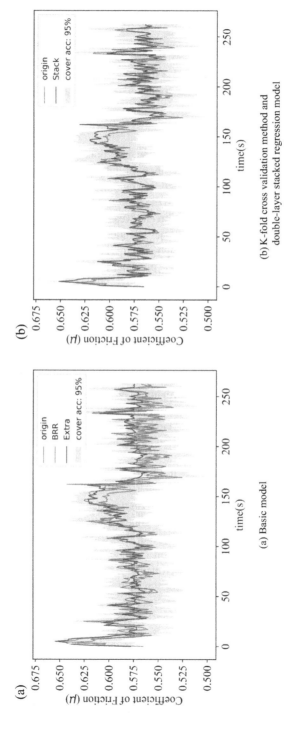

FIGURE 4.109 Comparison of the fitting results of the high load (20N) group with image information.

multi-dimensional friction information. The K-fold cross-validation method was used to fully utilize the friction information cross-section dataset, and evaluation indicators suitable for the friction coefficient fitting problem were defined. Stable and accurate friction coefficient fitting was achieved under multiple working condition friction information data samples, This provides a new technical method for real-time monitoring of friction coefficient.

4.3.5 Summary of The Case

Friction is a common physical phenomenon in production and daily life, and the friction information such as noise and vibration induced by friction has a significant impact on daily life, industrial production, equipment operation and maintenance, etc. Studying the friction information in the process of friction and wear, in order to achieve correlation monitoring of friction and wear status, is of great significance for improving living comfort, ensuring stable operation of equipment, and reducing the operation and maintenance costs of high-end equipment in complex environments.

Under the framework of friction informatics, the increasingly developed information technology has provided innovative ideas and methods for the study of friction information. Extensive research on various types of friction information has revealed the correlation characteristics between friction information and output variables such as friction coefficient. At the same time, it also faces the challenges of multi-dimensional friction information fusion, real-time collection of friction information, and data monitoring. Starting from the real-time data collection and processing of friction and wear tests, this chapter studies the correlation characteristics between friction information and output variables such as friction coefficient, achieving effective monitoring of friction and wear status, which has important academic value for the research and expansion of the field of tribology.

The full research work and main conclusions of the case are as follows:

1. Real-time collection method for friction and wear test data
   The collection of friction and wear test data is an important source of friction information research. Based on the Rtec-5000S friction and wear testing machine, this chapter integrates multiple sensing devices and builds a real-time collection system for friction information such as friction sound, friction vibration, and friction images. Through the collaborative thinking of experimental orientation and data collection methods, a real-time data collection method for radial vibration data of the upper sample fixture was proposed and practiced, solving the problems of unclear and easily obstructed real-time collection of friction images, and developing a new non-invasive method for researching friction information. On the basis of pretests targeting different materials and conditions of different durations, a test combination with more obvious friction behavior phenomena and more significant changes in friction information was selected. Through rigorous experimental plans and steps, real-time collection of friction and wear test data was achieved.

2. Multi-dimensional friction information feature extraction during friction process

Based on the multi-dimensional friction information raw data obtained from friction and wear tests, combined with the background knowledge of tribological behavior, data feature extraction with tribological research value was carried out. For various one-dimensional friction information, multiple methods have been used for feature extraction: for data such as friction coefficient and upper sample height, filtering methods have been used to highlight the main trend of change; extract envelope variation features that are in line with the characteristics of the experimental phenomenon based on the loading force data; extracting data features such as SPL and MFCC that are in line with human ear perception characteristics for acoustic information during friction and wear processes; and for friction vibration information, data features such as kurtosis and skewness that effectively reflect the trend of data fluctuation were extracted. For two-dimensional friction image information, image processing methods such as motion amplification algorithm based on deep learning were used to successfully extract friction information reflecting the radial small vibration of the upper sample, achieving data dimensionality reduction of image information, and expanding new ideas for friction process image research. By analyzing the changing characteristics and trends of multi-dimensional friction information, the correlation characteristics between friction information and friction coefficient are intuitively displayed, providing a basis for friction information correlation and driving friction data monitoring in complex scenarios.

3. An integrated regression model for fitting friction coefficient with multi-dimensional friction information

On the basis of multi-dimensional friction information feature time series data, a cross-sectional friction information dataset was formed through time dimension segmentation. The strong linear correlation between multi-dimensional friction information and friction coefficient was verified using methods such as correlation coefficient matrix and data distribution fitting, providing a data foundation for regression fitting of friction coefficient. In response to the problems of multiple friction information samples and complex patterns, the K-fold cross-validation method was used to improve the utilization of the dataset. Multiple basic models were selected to establish an integrated model for regression fitting of the time section friction coefficient, achieving fitting from multi-dimensional friction information to friction coefficient. The fitting accuracy under the CR index of the fitting results was stable at over 90%. The correlation between friction information and friction coefficient and its continuity in the time dimension have been proven using scientific methods, providing an effective solution for real-time monitoring of friction coefficient using friction information.

5 Conclusion

In order to improve the efficiency of tribological research and optimize the tribological research process, this book proposes the term tribological informatics. Tribology informatics improves the research efficiency and process of tribology by establishing tribological standards, establishing tribological databases, and utilizing information technology to collect, classify, store, retrieve, analyze, and disseminate tribological information.

We explained the conceptual framework and informational expression of tribo-informatics and introduced the application of tribo-informatics methods in tribology. Three case studies are offered, which might provide the readers with some ideas for applying tribo-informatics.

The application of information technology is able to significantly reduce the information entropy of the system, and the reduction of information entropy can improve the order of research information, thereby shortening research time. Therefore, the establishment and promotion of a complete tribology information system will inevitably shorten the research cycle of tribology and promote the research achievements of tribology more widely.

DOI: 10.1201/9781003467991-5

Index